Chemistry In a Day of Student's Life

Chemistry In a Day of Student's Life

Mahadev Kumbar

Writers Club Press
New York Lincoln Shanghai

Chemistry In a Day of Student's Life

Writers Club Press
an imprint of iUniverse, Inc.

For information address:
iUniverse
2021 Pine Lake Road, Suite 100
Lincoln, NE 68512
www.iuniverse.com

ISBN: 0-595-26512-X

Printed in the United States of America

To My Wife

Gangadevi

The Source of My Energy

Contents

Preface ix

Acknowledgements xi

Introduction xiii

The Plot xxiii

The Scenario 1

Appendix A: pH and Its Limitations 85

Appendix B: Fullerene I. Computer Simulation of C_{60} Molecule 93

Appendix C: Fullerene II. Computer Simulation of Clusters of C_{60} Molecules 135

Appendix D: Solution : Molecular Point of View and Molecular Weighing Balance 147

Appendix E: A Brief History of Chemistry 161

Appendix F: Frontiers in Chemistry 213

Index 227

About the Author 237

Preface

As a chemistry teacher, the most fundamental question, how can I make students understand chemistry? always haunts so deeply that some times you feel that you are left out in the cold. So, I decided to do something about it. What I had in mind was to teach students the importance and interaction of chemistry in their daily activities. My prime objective was to make students realize how much chemistry they come across in one day of their lives. Hence, I decided, as my first endeavor, to give students a project called **Chemistry in a day of student's life** in one of my general chemistry courses back in 1993.

Before I had assigned the topics, I had asked students about their interests and life goals. Based on their interests, I selected the topics and assigned to respective students, asked them to write 2-5 page term papers. As we know, merely writing term papers is not a good indication of understanding the subject matter. My main aim was to make them understand the subject matter they worked on. Hence, I had asked them to present their work in the classroom in front of their fellow students. This way they make, at least, a concerted effort to understand their assigned topics. This is a kind of **term paper-seminar concept.** In addition, I also had asked their fellow students to grade them based on their presentation and the quality of the material. I took their grading into consideration when I gave the final grade for the term paper. Based on their presentation, I developed the scenario to string their presented material in the form of **tale-tell** and further added more topics to complete the scenario.

Believe me! It was one of my exhilarating experiences that I ever had teaching chemistry. At the end of the semester, they were very happy and felt that they really understood chemistry. They also expressed that they learned more chemistry by doing this project than in any other chemistry course.

Subsequently, I repeated this experiment. The results were very similar. Thus the present book in part is the result of those chemistry courses. In this book, I have also added four research papers, a brief history of chemistry, and frontiers in chemistry in the form of appendices to compliment chemistry topics presented in the scenario.

Such projects, in my opinion, are only possible if the classroom size is small. As well, I also believe that it all depends upon the student population and motivation. At that time, all my students were graduates and coming back to school to change their careers. They were more matured and willing to learn.

One should not construe that this is the only scenario that revolves around student's daily life. One can, of course, design other types of scenarios based on students' interests. In any event, I feel that these kinds of projects **based on students' interests and goals rather than on instructor's interests and goals,** are certainly rewarding. Besides, such projects provide an opportunity to elevate the interaction between students and instructors.

Mahadev Kumbar
Plainview, New York
January 2003

Acknowledgements

I would like to thank my former students, especially, Melissa Agrusa, Jennifer Antes, John Brocks, Arthur Burdette, Celeste Eagleston, Peter Gold, James Johnson, Yukari Mitomo, Joi Pacchiano, Dina Pitkiewicz, Andrew Puch, Anne Sheridan, and Carlos Vieira for their participation in my first project and making that project a very successful one setting the ground for future projects.

Introduction

a. What is Chemistry?

b. Branches of Chemistry

c. Why Study Chemistry

d. Chemical Industries Serving Mankind

e. How to Read This Book

a. What is Chemistry?

Chemistry is the central and essential science, which can be defined in various ways: a science of atoms or molecules; or a science of reactions of different kinds of matter; or a science that reveals the inner beauty of molecular world; or a science that unravels the mystery of our nature. In essence, *the chemistry is the language of nature.*

The language of chemistry is very similar to any natural language. Thus, there exists a great similarity between the language of chemistry and the natural language including the process as well as the mechanics of learning.

In any natural language, the learning process usually takes place in step-wise starting with alphabets. Similarly, the learning process in chemistry also proceeds in step-wise starting with elements or atoms. The elements or atoms are considered to the building blocks of our nature as alphabets are to the natural language. The different kinds of symbols are assigned to alphabets for easy remembrance and quick recognition (for e.g., A, B, C, etc.). Likewise, the elements or atoms are also assigned numerous symbols for the same reason; each element carries a very distinct symbol (for e.g., H_2, O_2, etc.). The words, whether small or big, are created with various combinations of alphabets (for.

e.g. go, come, etc.). Likewise, molecules in chemistry, are also created by the combinations of elements or atoms (for e.g., H_2O, CO_2, etc.) The symbols for words are similar to the symbols for molecules, which are known as chemical formulas. The pronunciation of words is similar to the pronunciation of chemical formulas that are labeled as molecular names (for e.g., Dihydrogen oxide, Carbon dioxide, etc.). The words in natural languages are further strung into sentences of shorter or longer length using pre-defined set of rules known as grammar. The molecular names (chemical names) in chemistry are assigned using also a set of rules known as IUPAC (International Union of Pure and Applied Chemistry) nomenclature. The molecules in the language of chemistry are also further joined together to create bigger molecules. Eventually, many sentences in natural languages are used to build a paragraph or a page or a book. Likewise, molecules of smaller or bigger nature are used to build substances or bulk matter. This similarity is summarized below:

Natural Language	Language of Chemistry
Alphabets	Elements or atoms
Words	Molecules
Pronunciation	Molecular Formulas
Grammar	IUPAC Nomenclature
Sentences	Small, larger molecules
Paragraphs, pages or books	Substances, bulk matters

Chemistry is everywhere. Our very existence on this planet critically depends upon the science of chemistry. We encounter chemistry at each and every step of our lives without feeling its presence, yet we are so naive of its existence. The benefits it can offer are tremendous - **better chemistry for better life**. In this spirit, the scenario in this book is presented to illustrate " how chemistry is webbed all around our lives."

Chemistry is the first and foremost science. I believe that the creation and existence of matter is based on the laws of chemistry and the dynamics of the matter is governed by the laws of physics and mathematics. The laws physics or the mathematics cannot act on void, but they have to act on something. That something is the matter or chemical composition, which is the work of chemistry. Hence, the laws of chemistry, in my opinion, are more fundamental than any other laws even beyond the realm of quarks and leptons.

b. Branches of Chemistry

Chemistry is a branch of science concerned with creation, composition, properties, and structure of substances, and the transformation they undergo when they react or combine with other substances. More precisely, the chemistry can be defined as the science of atoms and molecules, and the interaction between them.

Chemistry can be dissected into divisions or branches according to the nature of substances studied or the types of studies undertaken. Organic chemistry and Inorganic chemistry fall in the first category and physical and analytical chemistry fall in the second category. The original distinction between organic chemistry and inorganic chemistry was based on the realization that compounds of biological origin were quite distinct from those of mineral origin in term of their overall properties.

The division of chemistry into organic, inorganic, and physical chemistry arose in the nineteenth century. Analytical chemistry came into existence in the twentieth century as a result of sophisticated and powerful instrumental techniques of analysis. During the same time the kinship between the biology and chemistry grew much closer, as a result of which biochemistry was born. Thus the main five braches of chemistry are,

- Inorganic chemistry
- Organic chemistry
- Physical chemistry

- Analytical chemistry
- Biochemistry

Inorganic chemistry is the study of elements (metals, non-metal, and rare gases). It investigates the structure and characteristics of non-living matter - minerals found in earth's crust as well as in the crusts of other planets or moons or inorganic matter of extraterrestrial nature. Branches of inorganic chemistry include synthetic inorganic chemistry, solid-state chemistry, coordinate chemistry, geochemistry, organometallic compound chemistry, reaction kinetics and mechanisms, nuclear science and energy, inorganic technology, crystallography, etc.

Organic chemistry is the study of carbon compounds is foremost highly investigated field of science. It is of vital importance in pharmaceutical, petrochemical, and textile industries, where its prime directive is the synthesis of new organic molecules, drugs, and polymers. Branches of organic chemistry include synthetic organic chemistry, physical organic chemistry, natural products, bioorganic chemistry, carbohydrates, polymers and polypeptides, structural elucidation, etc.

Physical chemistry is the study of physical properties of matter, for which it utilizes the concepts and techniques of physics and mathematics. In other words, the physical chemistry is an application of physics and mathematics to chemistry. Various branches of physical chemistry are, quantum chemistry, chemical physics, thermodynamics, chemical kinetics, electrochemistry, nuclear chemistry, polymer chemistry, solution chemistry, colloidal chemistry, structural chemistry, etc.

Analytical chemistry is the study of analysis of composition of a given sample of material. The analyses are of two kinds, qualitative and quantitative. In the qualitative analysis, the presence or absence of substances is identified whereas in quantitative analysis the exact quantities of substances are determined. The branch of analytical chemistry is the stoichiometry that deals with exact masses (weights) of chemicals participating in chemical reactions.

Biochemistry is the latest main branch of chemistry that undertakes the study of living systems, which is formulated as a fusion between biology and chemistry. Various braches of biochemistry are nucleic acids, enzymes, proteins and polypeptides, hormones, lipids and membranes, bioenergies, etc.

c. Why Study Chemistry?

Look around yourself, the food you eat is nothing but chemicals, the water you drink is a compound of two elements hydrogen and oxygen, the air you breathe is a mixture of various gases (gas is one of the physical states of matter, the other two are solid and liquid), the clothes you wear-cotton or synthetic- are nothing but natural fibers or synthetic fibers (nylon, polyester, etc), the medications you take are processed and manufactured by chemical means using various chemicals, the vehicle you drive is full of chemistry, the comforts you enjoy in and around your house are loaded with chemistry. In addition, most other aspects of your life are not only directly related to the chemistry but also are based on chemical principles and *know how*.

The most important thing in life is to make life more enjoyable, live longer and healthier, and make this world a better place to live. If you understand the relationship between chemical substances and the life, and how it affects the life cycle, then you understand and enjoy your life better. For example, if you know a particular substance is a carcinogen (cancer causing substance or agent), then you will watch out for that particular substance and will not buy any grocery containing that substance. This way, you not only keep yourself healthier but also become an educated consumer. Another example is smoking. If you know the harmful effects of chemicals present in cigarettes upon your health, then you will not smoke. You smoke because of your ignorance about the chemical substances present in cigarettes. Still another example would be the use of pesticides in and around your home. If you know the health hazards involved in using such pesticides and their effect on the ground water as well as on various ecological systems, then you will tend to avoid using them.

Thus, the understanding of chemistry not only makes you to understand your life better but also helps you to develop *logical reasoning* by analysis, which is also an important aspect of your life.

d. Chemical Industries Serving Mankind

- Biotechnology — Genetic engineering, gene therapy, genetic manipulation, producing new and effective drugs, new kinds of crops, etc.
- Brewing — Beers, light-beers, lager beers, etc.
- Cement — Variety kinds of cements for roads, pavements, driveways, buildings, etc.
- Chemical — Chemicals of all kinds needed in schools, colleges, universities, basic and applied research institutions, and industrial research, etc.
- Coal and tar — Extraction of coals, tars for roads, home, etc.
- Cosmetic — Various perfumes that include purely chemical based, natural extract based, animal extract based, coloring agents, face creams, face foundations, lipsticks, etc.
- Electrochemical — Batteries for cars, trucks, airplanes, radios, watches, hearing aids, electroplating, jewelries, silverwares, etc.

• Electronic, semiconductors	Television colors, LCD (Liquid Crystal Displays), computer chips, hard drives, sound cards, video cards, transistors, integrated circuits, semiconductors, etc.
• Energy producing	Nuclear, solar, thermal, electrochemical, wind, etc.
• Fertilizer	Different kinds of fertilizers for farms, lawns, gardens, vegetables, etc.
• Food, juices, soft drink	Solid foods, liquid foods, dry foods, canned foods, frozen foods, baby foods, dairy products, etc.
• Glass, ceramic	Automobile glass, stained glasses, glass wares for home use, commercially useable glasses, ceramic tiles, ceramic containers, etc.
• Industrial gas	Various gases including refrigeration gases like freons, etc.
• Industrial waste treatment	Converting harmful chemicals into harmless chemicals; nuclear, chemical, sewage, etc.
• Leather, tanning	Luggages, handbags, shoes, belts, attaches, leather wares, household goods, etc.
• Liquor	Hard liquors like Scotch, Vodka, etc.
• Lumbar	Building materials, furnitures, etc.
• Metallurgy	Metals, alloys, extraction of pure metals from ores, etc.
• Natural gas	Extraction of natural gas, etc.
• Nuclear	Nuclear plants, nuclear medicine, etc.
• Oil	Cooking oils, vegetable oils, etc.
• Paints, varnishes	Oil based paints, water based paints, varnishes, etc.

- Paper   Writing papers, envelops, paper towels, etc.
- Petroleum   Extraction and purification of crude oil, home heating oil, gasoline of various grades, etc.
- Petrochemical   Produce various chemicals based on the petroleum, etc.
- Pharmaceutical   Drugs, medications, vaccines, etc.
- Polymer, plastics   Building materials, household goods, etc.
- Pyrotechnology   Fire works, flares, fire crackers, etc.
- Rubber and tire   Tires for cars, trucks, airplanes, various rubber products, etc.
- Soap   Solid soaps, liquid soaps, detergents, bleaches, shampoos, conditioners, etc.
- Steel   Steels for homes, buildings, automobiles, railroads, etc.
- Sugar   White and brown sugars, cane sugar, beet sugar, molasses, etc.
- Tobacco   Cigarettes, cigars, etc.
- Vitamins   Different kinds of vitamins for consumer needs, etc.
- Water purification   Drinking water, bottled waters, etc.
- Winemaking   Table wines of all kinds, champagnes, etc.

Please note that this list is by no means complete.

e. How to Read This Book

The book is presented in such a way that each step in the scenario is followed by relevant chemistry topic or topics. I advise you to read first only the plot and the scenario (these paragraphs are in italic bold face) that surrounds the student so that you can understand the flow of logic and get the feeling for the book. Once you have read the scenario, then go back and start reading from the beginning, this time the scenario with chemistry part under the heading " Chemistry in Service."

The Plot

This is a usual day in the daily life of a fictitious student by the name Mercury, who is a liberal arts student at Nassau Community College; a two-year liberal arts school located about 25 miles east of New York City on Long Island. Mercury lives home, which is about 10 miles away from the school, and drives the car to commute between school and her home. Even though, Mercury is a liberal arts student, she is very much interested in science, and especially in chemistry, because she knows for sure that chemistry is deeply embedded in the life cycle on this planet. She belongs to a new generation, which profoundly concerns with health, environment, and the resources it consumes including food. So the following scenario explains step-wise about chemistry in one day of Mercury's life.

The Scenario

Every day, five days a week - rain or shine - Mercury wakes up early in the morning due to her early classes, and also due to the fact that she needs some time to finish up her morning activities, before she can head-on to the school. Mercury being a very conscious person, she does not want to be labeled as halitosis. Therefore, as soon as she opens up her eyes, she walks into the bathroom and grabs the mouthwash bottle, and gargles with about 10-15 ml of mouthwash for about 2-3 minutes.

Chemistry in Service

Mouthwash: Mouthwash is a homogeneous solution containing various chemicals coexisting in a peaceful manner. The mouthwash acts as an antiseptic - kills microorganisms that cause bad breath, plaque and gingivitis. The ingredients in a typical anti-plaque mouthwash are the followings: glycerin, alcohol, sodium benzoate, sodium bicarbonate, sodium salicylate, and sodium borate.

Glycerin ($C_3 H_5 (OH)_3$) is derived from fats and oils is used as a moistening and sweetening agent. Ethanol (C_2H_5OH) is used for its solubility factor. Benzoic acid (C_6H_5COOH) (insoluble in water) in the form of sodium benzoate ($C_6H_5COO^-Na^+$)(soluble in water) is used as a preservative, which has the ability to retard the rate of spoilage. Salicylic acid ($C_6H_5 (OH) COOH$) in the form of sodium salicylate

$(C_6H_5(OH)COO^-Na^+)$ is used as an analgesic and also as a preservative. And sodium bicarbonate $(NaHCO_3)$ commonly known as the baking soda is used as a base, which has a soft and soapy character.

When sodium bicarbonate comes in contact with an acid, such as, sulfuric acid that is formed by bacteria, the chemical process known as **effervescence** takes place due to the evolution of carbon dioxide gas according to the following equation:

$$2NaHCO_3 + H_2SO_4 \quad \rightarrow \quad Na_2SO_4 + 2H_2CO_3 \quad \rightarrow \quad Na_2SO_4 + 2H_2O + 2CO_2(g)$$

The process is nothing but neutralizing the acid content of the mouth.

The percentage of the active ingredients is quite small compared to water, which is about 92 % of the total volume. Alcohol is present about 7.5%-8.0 % of the total volume that leaves about 0.5% for the remaining active ingredients. The use of approximately 30 ml. of rinse costs about 12 cents, a small price for such huge benefits.

In addition to gargling, it is equally important for Mercury to brush her teeth with toothpaste, not only to eliminate the mouthwash flavor but also to expel any leftover residues. Hence, Mercury decides to brush her teeth with a brush and tarter controlling toothpaste.

Chemistry in Service

Toothpaste: - Toothpaste is a homogeneous paste (semi-solid). It is quite important to remove any starch and sugar from teeth because these are the adored foods for bacteria and are broken down by bacteria into acids. These acids in turn **demineralize** teeth. Saliva is a natural fighter against bacteria, in that, it helps to neutralize the acids formed by bacteria with its inorganic buffering compounds. Saliva is a weak acid (pH ~ 6.5-7.2) and mostly water.

Acids and bases are very important class of substances occurring in nature. Acids are identified by the presence of proton (H^+) or hydronium ion (H_3O^+), and bases by the presence of hydroxide (OH^-) ion. They are categorized by the range of pH. A detail account of pH may be found in Appendix A. Some common properties of acids are; (a) sour taste, (b) neutralize bases, (c) pH lower than 7, and (d) turn blue litmus paper into red. Some common properties of bases are;(a) bitter taste, (b) slippery feeling, (c) neutralize acids, (d) pH is greater than 7 , and (e) turn red litmus paper into blue.

Cavities form when saliva cannot supply calcium and phosphate fast enough to offset demineralization. Salivary amylase or Ptyalin hydrolyses boiled starch into dextrin and maltose. Two of the bacteria formed are *streptococcus mutans, which* grows on sugar and *actinomyces*, which ferments any carbohydrate into acids. These acids in turn break down the tooth enamel, which is made up of a small layer of hydroxyapatite, $Ca_5(PO_4)_3$ OH. The demineralization process is to break down of this substance to release calcium ion, phosphate ion and hydroxide ion:

$$Ca_5(PO_4)_3\,OH \quad \xrightarrow{\text{Demineralization}} \quad 5\,Ca^{2+} + 3\,PO_4^{3-} + OH^-$$

Thus, the released hydroxide ion (base) dissolves tooth enamel.

Brushing with toothpaste helps to prevent some of these problems. Toothpaste basically has two actions- detergent action and abrasive action. The substance used as a detergent suspends food particles in water so they may be rinsed away. The substance used as abrasive cuts into surface deposits. Some common abrasives are calcium carbonate ($CaCO_3$), hydrated silica - a type of sand, and hydrated alumina. The ingredients vary from brand to brand. Some common ingredients are:

monofluorophosphate (PO_3F^{2-}) (MFP), sodium phosphate (Na_3PO_4), hydrated silica and sorbitol. Monofluorophosphate contains fluoride ion (F^-) which has the ability to inhibit certain enzymes, such as, those that act as catalysts in the fermentation of carbohydrates to produce lactic acid. Fluorine, a halogen, is the most powerful of all chemical-oxidizing agents. Fluoride ions are essential in the structural formation of teeth and are released through the following chemical reaction:

$$PO_3F^{2-} + H_2O \rightarrow H_2PO_4^- + F^-$$

Fluoride ions substitute into the hydroxyapatite when **remineralization** occurs -hydroxide ions are replaced by fluoride ions.

$$5\ Ca^{2+} + 3PO_4^{3-} + (1-x)\ OH^- + x\ F^- \rightarrow Ca_5(PO_4)_3(OH)_{1-x}F_x \quad ; x = 0 \text{ or } 1$$
$$\text{reminerlization} \quad \text{(fluoridated hydroxyapatite)}$$

This is possible because of the similarity in size of fluoride ion and hydroxide ion. Some of the hydroxide ions of hydroxyapatite structure are replaced by the fluoride ions, but the sum of the number of hydroxyapatite ions (1-x) and the number of fluoride ions (x) equals 1 for every 5 calcium and 3 phosphate ions in the enamel. As more Ca^{2+} and PO_4^{3-} become available via demineralization and also from saliva, the remineralization process is increasingly favored. If free F^- is available it becomes incorporated into the hydroxyapatite structure.

The typical toothpaste contains, sodium phosphate (Na_3PO_4 .$12H_2O$) -used to soften hard water which destroys soap action, hydrated silica (SiO_2), a form of sand- used as an abrasive to cut into food and calcium deposits, and sorbitol ($C_6H_{14}O_6$), a sweetener - used to help prevent bacteria from sticking to enamel.

A very thorough study conducted by United Kingdom Royal College of Physicians concluded that there is no hazard to individuals or environment from fluoride levels up to 1 ppm that is greater than the levels used in drinking water. 100 ppm can cause damage to teeth in a condition known as *chronic endemic dental fluorosis* but only in developing teeth.

After finish brushing her teeth, Mercury decides to take a shower. She gets into the bathtub, turns the shower on for few minutes, and applies soap to her body to clean it thoroughly. She starts feeling clean and fresh as soap gets down to the business of cleansing.

Chemistry in Service

Soaps: When fats are heated with aqueous solutions of bases, glycerol and salts of fatty acids ($CH_3 (CH_2)_n COO^- M^+$, n=12 to 18) are formed. This process is known as saponification meaning soap making. The type of soap obtained depends upon the nature of the base used in saponification process - a solid soap consists of sodium salt of fatty acid and a liquid soap involves potassium salt of fatty acid.

$$
\begin{array}{lll}
CH_2 - O - CO - (CH_2)_n - CH_3 & & CH_2 - OH \\
| & & | \\
CH - O - CO - (CH_2)_n - CH_3 & + \ 3\,M^+OH^- \quad \rightarrow & CH - OH + \quad 3\,CH_3 - (CH_2)_n - COO^-M^+ \\
| & & | \\
CH_2 - O - CO - (CH_2)_n - CH_3 & \text{heat} & CH_2 - OH \\
\text{fatty acid(fat)} & \text{base} & \text{glycerol} \qquad \text{soap}
\end{array}
$$

The soap contains two parts; the hydrophobic (water hating) hydrocarbon chain ($CH_3-(CH_2)_n -$) and hydrophilic (water loving) head group ($-COO^- M^+$). Soap molecules do not exist separate from one another in solution. The hydrocarbon chain tends to associate together with other hydrocarbon chains forming a colloidal dispersion of spherical aggregates called *micelles*.

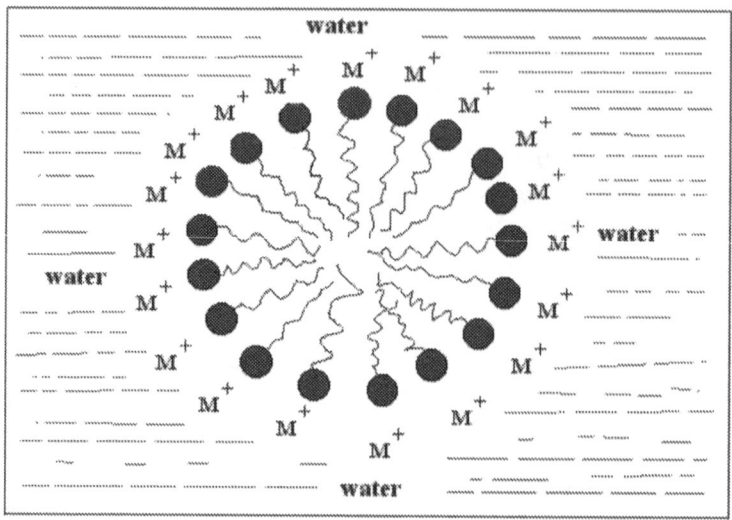

The hydrophilic salt-like ends are on the surface exposed to the water. The soap action is accompanied by dissolving the hydrocarbon chain in oil or grease (non-polar) and the head group in water (polar). This is to say that **like likes like**. The soap action leads to what is known as an emulsion. Emulsion is when one liquid is dissolved in another, in other words, when there is a stable suspension of one liquid in another. Dirt and oils become suspended within the soap. When dirt and oil are in suspension they are easily washed off. This is how soap works.

Soap has been in existence well over 2300 years, assumed to be known to ancient Celts and Romans. In spite of that it was only used in medicine as late as sixteenth and seventeenth century. But by the nineteenth century, it has come into widespread use. It is estimated that annual world production of ordinary soaps, not including synthetic detergents, is well over 6 million tons.

While Mercury deep into shower, notices that there is a foul filmy covering known as scum (scale) around the bathtub. She is now somewhat puzzled and begins to ponder herself, how this undesirable looking coat on the bathtub could have been formed? But she does not know that the scum is most likely the cause of the chemical reaction between the soap and the water, especially, hard water.

Chemistry in Service

Scum: The scale or soap scum, as generally known, is caused by the reaction of soap with hard water (the hard water is the water that contains heavy ions such as Fe^{3+}, Ca^{2+} and Mg^{2+}). These ions react with the soap by replacing Na^+ or K^+ ion and forming a precipitate known as the soap scum, which is left behind in the bathtub after the liquid has gone down the drain. The chemical reactions for the formation of scum are:

$$2\ CH_3(CH_2)_nCOO^- Na^+ + M^{2+} \rightarrow [CH_3(CH_2)_nCOO^-]_2 M^{2+} + 2\ Na^+\ ; M^{2+} = Ca^{2+}\ or\ Mg^{2+}$$

and

$$3\ CH_3(CH_2)_nCOO^- Na^+ + Fe^{3+} \rightarrow [CH_3(CH_2)_nCOO^-]_3 Fe^{3+} + 3\ Na^+$$

$$\text{soap} \qquad\qquad\qquad \text{scum}$$

After the shower, she dries herself with a bath towel made from 100% cotton and decides to dress up. It being a summer time, Mercury needs no heavy dresses. So, she puts on a jean made from 100% cotton, a shirt made out of mixtures of synthetic fibers, such as, nylon, polyester, and rayon. Mercury being a very down-to-earth person, she does not use any kind of hair sprays due to their role in the destruction of ozone layer or makeups due their chemical sensitivity.

Chemistry in Service

Cotton: Cotton, being a light and airy material, is the most important natural fiber belongs to the genus of *gossypium* and to the families of *mallvacea, herbsutum,* and *herbacem.* Since the prehistoric times, this celluloid fiber has been spun, woven, and dyed in Egypt, India, and China. It is also known that the cotton cloth was produced in Peru at least 100 years ago.

Synthetic Fibers: Synthetic fibers (man-made) are **polymers** - containing many monomers - are truly remarkable and illustrate chemists' craftsmanship. Since 1930, a large number of man-made polymers have been manufactured. They provide cheaper and better clothing than cotton due to inexpensive production. Cloths are usually made out of rayons, nylons, and polyesters.

Acetate rayon is produced when an acetate (CH_3COCH_3) solution of cellulose diacetate is passed through spinnerets and evaporated the solvent. Acetate rayons have poor strength but seem to possess a desirable feel.

The first truly synthetic fiber was **nylon**, which was first produced by W.H.Crothers of E.I.duPont de Nemours around 1930. There are of course several commercial nylons, but the original synthetic fiber nylon-66, is produced by heating adipoyl chloride and hexamethylene diamine producing the repeating unit known as polyamide:

$$Cl\text{-}CO\text{-}(CH_2)_4\text{-}CO\text{-}Cl + H_2N\text{-}(CH_2)_6\text{-}NH_2 \rightarrow Cl\text{-}CO\text{-}(CH_2)_4\text{-}CO\text{-}[NH\text{-}(CH_2)_6\text{-}NH\text{-}CO\text{-}(CH_2)_4\text{-}CO]_n\text{-}NH\text{-}(CH_2)_6\text{-}NH_2 +nHCl$$

| adipoyl chloride | hexamethylene diamine | repeating unit of nylon-66 |

A typical nylon-66 consists of at least 100 repeating units in one polymer molecule. The properties of various nylons depend upon the number of methylene ($-CH_2$) groups in the repeating unit. The strength or the tenacity of nylon-66 increases by stretching or drawing.

Much of the strength in the nylon is the result of hydrogen bonding between carbonyl (C=O) groups in one polymer chain and the hydrogen atoms of the amide groups in an adjacent polymer chain.

$$
\begin{array}{ccc}
| & & | \\
C = O & - - - - - - - - & H - N \\
| & & | \\
N - H & - - - - - - - - & O = C \\
| & & |
\end{array}
$$

The most widely used synthetic fiber is a **polyester** (polyethylene terphthalate), $-(O_2CC_6H_4CO_2C_2H_4)_n$ - , is synthesized by the reaction of terephthalic or its derivatives and ethylene glycol:

$$HO_2C-C_6H_4-CO_2H + OH-CH_2-CH_2-OH \rightarrow - (O_2C-C_6H_4-COOC_2H_4)_n$$

terephthalic acid   ethylene glycol   polyester

Garments made from polyesters have low moisture absorption with sticky feeling. However, when blended with cotton, the moisture is absorbed by eliminating undesirable effect by producing what is known as "wash-and-wear" wrinkle resistant garments. More than 2 millions tons are produced in USA. This product is known as **Dacron** in the United States, **Terylene** and **Crimplene** in the United Kingdom, and **Trevira** in Germany.

Acrylic fibers are produced by the polymerization of acrylonitrile ($CH_2 = CHCN$). They are produced by running a solution of the polymer in dimethylacetamide through spinnerets and finally evaporating solvents.

$$H_2C = CH\ (CN) \qquad \rightarrow \qquad -[-H_2C-CH(CN)-]_n-$$
$$\text{polymerize}$$

Polypropylene with repeating unit of $(-CH_2-CH-CH_3-)_n$ is another synthetic fiber, which is used in indoor-outdoor carpets and ropes.

$$H_2C = CH\ (CH_3) \qquad \rightarrow \qquad -[-H_2C-CH\ (CH_3)-]_n-$$
$$\text{polymerize}$$

Another synthetic fiber called **Spandex**, is a copolymer with polyester units and polyurethane, $-[(CH_2)_3\ NHCOO-]$, units. The flexible polyester units provide elasticity to the material, which is controlled by the not so flexible polyurethane units. These fibers have found their applications in foundation garments and swimsuits.

It is now time for Mercury to energize herself by eating a good and hearty breakfast. She is always in a hurry because she does not want to be late for the classes. Her daily breakfast consists of a glass of orange juice, a bowl of cereal and milk, and a cup of coffee.

Chemistry in Service

Orange juice: Orange juice is the natural product containing various health-important chemicals. Among them, the most prominent one is the vitamin C, also known as the L-ascorbic acid ($C_6H_8O_6$, melting point $= 192^0$ C, pleasant, sharp-acid taste) containing a five-membered unsaturated ring with two hydroxyl groups attached to two carbons which are doubly bonded.

The biosynthesis of ascorbic acid from D-glucose requires a specific enzyme, which is lacked in humans, guinea pigs, monkeys, and in a few vertebrates. Hence, ascorbic acid must be included in the diet. Its lack can cause *scurvy*, a disease that causes the weak blood vessels, loosening of teeth, lack of ability to heal wounds, hemorrhaging and in some cases eventual death.

Ascorbic acid also needed in the synthesis of collagen, a structural protein of skin, bone, connective tissue, tendon, and cartilage.

Can vitamin C help to cure the common cold? According to Linus Pauling it helps to prevent the cold in large doses. In any event, it is still a controversial matter. For some, it may be effective, for others, it may not. When one seems to be coming down with a cold, one needs at least two grams a day to head off the cold.

Milk: Milk is the natural fluid that contains proteins, carbohydrates, minerals, and fats. The milk provides about 150 calories per serving, 8 grams of protein, 11 grams of carbohydrates and 8 grams of fat.

The energy content of the food is measured in terms of calories [The calorie (cal) is defined as the amount of heat needed to raise the temperature of one gram of water by one degree centigrade. The kilocalorie (kcal) is 1000 cal, a much larger unit. In chemistry, both these units are in usage. The calorie used as the nutritional value, however, is not a calorie as defined above, but it is equivalent to kcal. Therefore, one nutritional cal or some times it is labeled as Cal (big C) is equal to one kcal].

Cereal: Cereal, like milk, is also a natural food that also contains carbo-hydrates, fat, cholesterol, sodium and potassium, vitamins, minerals and fiber, and provides with about 110 calories per serving. Proteins are part of every cell in the body. They serve as regulatory functions as well as structural functions. Carbohydrates occur in plants and animals and are quite essential for life. They include cellulose, starch and sugars. They are classified into three groups according to their structure, monosaccharides, disaccharides and polysaccharides, and are used by plants and animals to store glucose, which is the source of food and energy. Also they provide some mechanical structures of cells. Polysaccharides are stored as food energy. The most common starch found in plants is amylopectin. Cellulose is the most abundant struc-tural polysaccharide consists of glucose molecules. There is an abun-dance of this on the walls of plants.

Coffee: Coffee is produced from the roasted seeds of *coffea arabic* and contains a prime substance known as the caffeine (molecular formula $C_8H_{10}O_2N_4$), which has the following structural formula.

Caffeine causes the stimulation of the central nervous system (CNS) with increased wakefulness, increased heartbeat and metabolic rate. Caffeine does this by inhibiting an enzyme called *phosphodiesterase*, which is known to inactivate cyclic-AMP (Adenosine 3,5-cyclic monophosphate), a molecule that participates in an energy supply pathway. There could also be an increased capacity for physical work and diuresis. Secretion of acid in the stomach can also accrue as results of drinking the coffee. Coffee and other caffeine filled drinks contain another chemical known as the methylxanthines. Complete avoidance of methylxanthines may help control breast disease and birth defects, but there is no clear evidence to that effect.

After finishing her breakfast, Mercury puts on her shoes and ready to go to school. Grabs her knapsack and starts walking towards her car. As she approaches the car she suddenly realizes that the slick-looking car of hers is full of chemistry: body is coated with a paint; some of the interior parts, like dash-board, etc. are made from polymer material; windshield is nothing but a glass; battery uses oxidation-reduction principle; chrome fender on her car is created using the electrolysis process; tires are manufactured from rubber; air in the tires is nothing but a mixture of gases; gasoline in the engine burns using combustion principle; car moves because it uses the laws of thermodynamics; and catalytic con-verter is there to reduce air pollution and etc. When she gets closer to the car she notices that a spot of rust, from nowhere, has appeared on her car's chrome fender without giving any warnings.

Chemistry in Service

Paint: The paint on the car has two purposes, decorative and protective. Paint is a mixture of a liquid and one or more colorants known as pig-ments. The liquid is called adhesive (binder) consists of a solvent or thinner and the coating agent. The colored powders that impart colors are known as pigments. Pigments are two types - prime and inert. Prime pigments give color to paints. They may be inorganic substances: titanium dioxide (TiO_2) for white paints; iron oxides (FeO, Fe_2O_3, and Fe_3O_4) for red, brown, and yellow paints; organic substances, such as phthalocyanine for blue and green paints. Inert pigments like calcium carbonate ($CaCO_3$), talc, clay, and magnesium silicate are added to make the paint last longer. In addition, the paint may contain also some other special agent or additive to perform a special function.

The adhesives consist of water, resins, and liquids such as natural or modified natural oils. For example, latex adhesive is made by suspending synthetic resin like poly (methyl methylcrylate) in water. This suspension is an emulsion and the paints are known as emulsion paints but generally known as latex paints. When adhesive comes in contact with air, it evaporates (dries out) leaving behind a solid coating - in latex paints, the water evaporates leaving behind a film of resin. There are various types of paints formulated for specific purpose.

Latex paints. These are polymer latexes to which pigments have been added. The paint film is formed by coalescence of polymer molecules upon evaporation of water. These paints are water soluble, though the polymer in the paint is not, and due to that they are easy cleanup. In addition, they are quick drying and have little odor.

Oil paints: Suspension of pigments in oil such as linseed oil. The film is formed by the reaction of paint with oxygen when atmospheric oxygen polymerizes and cross-links the drying oil. In some instances, catalysts are added to accelerate the cross-linking process. Oil paints, once dried (cured, cross-linked) become no longer soluble, but can be removed through polymer degradation using paint strippers.

Oil varnishes: Varnishes consist of polymers, natural or synthetic, dissolved in drying oils with appropriate additives (catalysts) to speed up the cross-linking with atmospheric oxygen. When dried, they produce a clear and tough film.

Enamels: Enamels in fact are oil varnishes with pigments added. The added polymers are selected in such a way as to provide glossier and harder coating than oil varnishes.

Lacquers: These are polymer solutions to which pigments are added. The film, the polymer coating, is formed when solvent evaporates. The film is formed without cross-linking with oxygen and hence their surface exhibits poor resistance to some organic solvents.

Glass: Glass is defined as a rigid liquid of an inorganic product of fusion that has been cooled down to a rigid condition without crystallization. Glass is mainly silica sand (SiO_2) and made by heating silica sand and additives in a special manner; the raw materials are heated to high temperature in a furnace. A cullet, recycled or waste glass (5 to 40%) is added to the principle raw material to assist in the fluxing and melting. When the mass is fused some time is allowed for the gases to escape, and then the liquid is withdrawn and worked. While in hot condition, it is blown, molded, or rolled to give a desired shape, which is eventually retained when cooled. The range of temperature through which the glass is workable and its viscosity are important factors, which depend upon the amount of silica, nature of additives, etc. The common glass is essentially a sodium silicate in composition, it is 95 to 99% silica sand with the remainder being soda ash (Na_2CO_3) or salt cake (Na_2SO_4), lime stone ($CaCO_3$) or burnt lime (CaO), feldspar ($R_2OAl_2O_3\ 6SiO_2$), boric acid (H_3BO_3) or borax ($Na_2B_4O_7.10H_2O$) along with the appropriate coloring or discoloring agents, reducing (for e.g., carbon is used to reduce the sulfate) or oxidizing agents, and crushed cullet. When heated, all these substances form oxides and get integrated into the silicon structure.

$$Na_2CO_3 \quad \rightarrow \quad Na_2O \quad + \quad CO_2(g)$$

soda ash (sodium carbonate)   sodium oxide   carbon dioxide

| CaCO$_3$ | $\rightarrow$ | CaO | + | CO$_2$(gas) |
| limestone(calcium carbonate) | | calcium oxide | | carbon dioxide |

| R$_2$OAl$_2$O$_3$6SiO$_2$ | $\rightarrow$ | R$_2$O | + | Al$_2$O$_3$ | + | 6SiO$_2$ |
| feldspar | | | | alumina(aluminum oxide) | | |

| 2H$_3$BO$_3$ | $\rightarrow$ | B$_2$O$_3$ | + | 3H$_2$O |
| boric acid | | boron oxide | | |

The exact structure of the glass depends upon the ingredients and the processing conditions.

Colored glasses have been known as early as to Egyptian civilization and later to Roman civilization. To produce colored glasses small amount of the desired coloring agent is usually added to the charge: cobalt oxide (CoO) for blue glass; Manganese oxide (MnO$_2$) for violet; Gold oxide (Au$_2$O$_3$) or selenium oxide(SeO) for red; Nickel oxide (NiO) for yellow to purple; calcium fluoride (CaF$_2$) or arsenic trioxide (AsO$_3$) or aluminum oxide (Al$_2$O$_3$) or zinc oxide (ZnO) or calcium phosphate (Ca$_3$(PO$_4$)$_2$) for opalescent white color; ferric oxides (Fe$_2$O$_3$) or salts for green; ferrous oxides (FeO) or salts for yellow; cupric oxide(CuO) for red, blue, or green; stannic oxide (SnO$_2$) for opaque; and iridium oxide(IrO$_2$) or mixture of other oxides for black color.

Battery: Battery is a device where chemical energy is converted into electrical energy when a spontaneous chemical reaction is carried out. The electrons are produced and drawn into the external wire to perform useful work. The battery used in the automobiles is known as the lead storage battery is composed of six galvanic cells, each with capacity of producing 2 V, that are connected in series to produce a total of 12 V. A typical automobile battery is shown below.

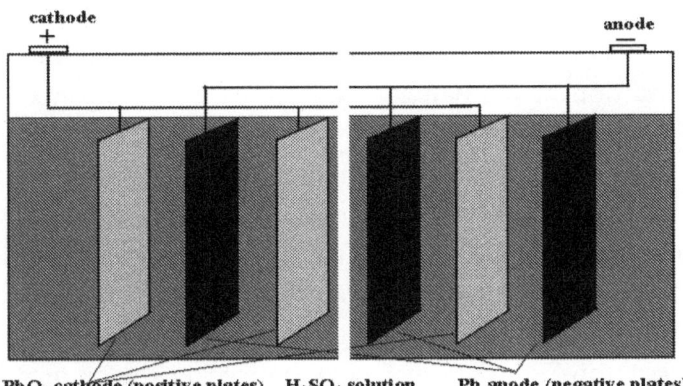

PbO₂ cathode (positive plates)　H₂SO₄ solution　　Pb anode (negative plates)

In each cell, the anode consists of lead plate and cathode composed of another set of plates that hold a coating of lead (IV) oxide (PbO_2). The electrolyte is a concentrated sulfuric acid (H_2SO_4). The battery makes use of oxidation-reduction reaction. When the battery is discharging (converting stored chemical energy into electrical energy), the following redox reactions take place at anode and cathode:

Anode reaction

$$Pb(s) + SO_4^{2-} (aq) \quad \rightarrow \quad PbSO_4(s) + 2e^-, \qquad \text{oxidation}$$

Cathode reaction

$$PbO_2(s) + 4H^+(aq) + SO_4^{2-}(aq) + 2e^- \quad \rightarrow \quad PbSO_4(s) + 2H_2O(l), \qquad \text{reduction}$$

The electrons thus produced in the anode reaction climb up the anode electrode, run through the external wire to run the engine, and come back down the cathode electrode and reduce the lead oxide. The overall reaction-taking place in each cell when the battery is discharging is sum of above two reactions:

$$PbO_2(s) + Pb(s) + 4H^+(aq) + 2SO_4^{2-}(aq) \quad \rightarrow \quad 2PbSO_4(s) + 2H_2O(l)$$

As the battery discharges the concentration of the sulfuric acid decreases. This phenomenon provides a convenient method for checking the condition of the battery. The decrease in the concentration of the sulfuric acid is monitored by measuring the density with a device known as hydrometer (this instrument consists of a rubber bulb on the top which is used to draw the battery fluid into to glass tube containing a float, the narrow neck of the float is usually marked in color to indicate the state of the battery and also ease of reading). The level to which the float sinks or rises depends upon the density of the liquid - lower the sink float, lower the density meaning weaker the battery. The great advantage of the lead storage battery is that electrolytes can be restored to their original or near original conditions. This is possible because applying voltage from external electrical source can reverse the above cell reactions occurring spontaneously during discharge. This is an electrolysis process where electrical energy is converted back into chemical energy. The disadvantage of the lead storage battery is that it is corrosive due to sulfuric acid and heavy to carry around.

Electrolysis: Electrolysis is the process of using electrical energy to make the nonspontaneous chemical reaction to occur in electrolytic cell. It is commercially exploited in the process known as electroplating, for example, silver and gold plating, etc. The electrical energy is usually supplied from the external source. The schematic representation of the electrolytic cell is shown below:

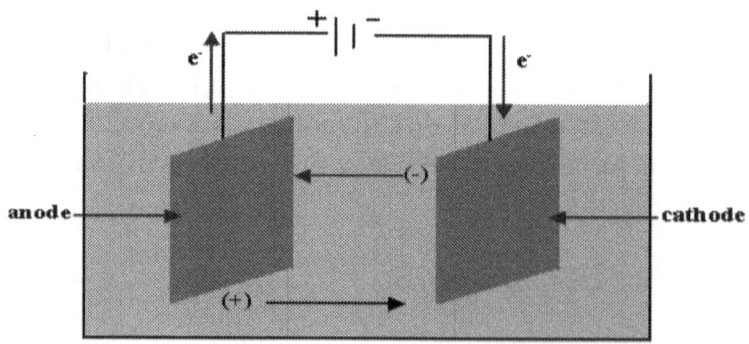

The cell consists of two electrodes, anode and cathode, immersed in a solution containing a metal ion, M^+ and a negative ion X^-. The metal ion is the desired ion that is to be coated on a specific surface, for example, chromium ion, in chromium bumpers on the car. The external course of electricity acts as an electron pump, pushing electrons into the cathode, and removing them from the anode. At the anode, the electrons are generated by the oxidation of an ion or molecule, while at the cathode, the metal ion undergoes reduction by accepting electrons. The reactions at these electrodes are:

Anode (oxidation reaction): $\quad 2X^- \quad \rightarrow \quad X_2 + 2e^-$

Cathode (reduction reaction): $\quad 2M^+ + 2e^- \rightarrow \quad M(s)$

Thus the metal produced at cathode gets deposited on the surface of the cathode.

Rubber: It is called rubber simply because it can be used to rub off pencil marks, is an elastic polymer. Natural rubber is commercially prepared from the coagulated latex of rubber tree, *Hevea brasiliensis*. Rubber is brittle and hard when cold, and gluey and soft when warm. In order to improve its elastic properties, Charles Goodyear invented a process in 1839, which is known as the vulcanization process, where rubber is heated with sulfur. The styrene-butadiene rubber (SBR), the copolymer obtained from the mixture of styrene and butadiene which further vulcanized and used in making the automobile tires.

$$n\, CH_2 = CH\, (C_6H_5) + 3n\, CH_2 = CH\text{-}\, CH= CH_2 \quad \rightarrow \quad -[\,-\,CH(C_6H_5) - CH_2\text{-}CH_2\text{-}CH=CH\text{-}CH_2\text{-}]_n$$

styrene $\qquad\qquad\qquad\qquad\qquad$ butadiene $\qquad\qquad\qquad\qquad\qquad\qquad\qquad$ SBR

Catalytic Converter: When the car is run some desirable chemicals (carbon dioxide) and some not so desirable chemicals (pollutants) are produced and discharged into the air. The catalytic converter is designed to reduce the concentrations of air pollutants in the exhaust that come out in the muffler when the car is run. The pollutants are basically carbon monoxide (CO), unburned hydrocarbons (HC), and nitrogen oxide (NO). The catalytic converter is equipped with a catalytic converter [contains platinum (Pt), palladium (Pd) or transition metal oxide such as copper (I) oxide (Cu_2O) or chromium (III) oxide (Cr_2O_3)] that facilitates the break down and conversion of these pollutants into harmless chemicals. Air is mixed with the exhaust jet and passed over catalyst in a converter that absorbs CO, NO, and O_2. The CO is oxidized to CO_2 by combining with oxygen, the NO is converted to N_2, and O_2 and unburned hydrocarbons are oxidized to CO_2 and H_2O:

$$2CO(g) \quad + \quad O_2(g) \quad \rightarrow \quad 2CO_2(g)$$
$$2NO(g) \quad \rightarrow \quad N_2(g) + \quad O_2(g)$$
$$2C_nH_{2n+2}(g) + (3n+1)\,O_2(g) \quad \rightarrow \quad (2n)\,CO_2(g) \quad + (\,2n+2)H_2O(g)$$

Unfortunately, there is a draw back in using the catalytic converter. All fossil fuels contain sulfur as impurities. The sulfur undergoes oxidation to sulfur dioxide (SO_2) during combustion. The sulfur dioxide itself is an air-pollutant, undergoes further oxidation to sulfur trioxide (SO_3), which is the worst pollutant. These two react with water vapor in the air producing what is known as acid rain that is nothing but sulfuric acid and/or sulfurous acid.

$$2S(g) \quad + \quad 2\,O_2(g) \quad \rightarrow \quad 2\,SO_2(g)$$
$$SO_2(g) + \quad H_2O\,(aq) \quad \rightarrow \quad H_2SO_3\,(aq)\ (\text{sulfurous acid})$$

and
$$2SO_2(g) + \quad 3O_2(g) \quad \rightarrow \quad 2SO_3(g)$$
$$SO_3(g) \quad + \quad H_2O\,(aq) \quad \rightarrow \quad H_2SO_4\,(aq)\ (\text{sulfuric acid})$$

Acid rain has devastating effect on our environment; ruins the beauty of our national monuments, pollutes lakes affecting aquatic life, destroys crops, and forests, etc.

Rust: Rust is a brown colored film coated on a metallic surface, usually iron metal, such as iron nail. It is a form of an oxide that is created when an iron (Fe) reacts with the atmospheric oxygen (O_2) in the presence of moisture (H_2O) (present also in the air). This process can be described by the following oxidation-reduction (Redox) reaction:

$$4Fe\,(s) + 3O_2\,(g) + nH_2O \quad \rightarrow \quad 2\,Fe_2O_3.nH_2O$$

In this reaction, Fe is oxidized and O_2 is reduced (oxidation is the process of loosing electrons, and reduction is the process of gaining electrons).

Mercury now opens the car door and sits on a driver side and ready to start the car. As she sits down on the vinyl seat, she feels that the vinyl seat is somewhat warmer than usual, as the car has been standing in the morning sun for awhile. She becomes curious as to the nature of the vinyl seats.

Chemistry in Service

Vinyl seats: Most cars are equipped with the vinyl seats because they are cheaper. The inlay is a long chain synthetic polymer known as polyvinyl chloride (PVC) with the following chemical structure.

$$H_2C = CH\,(Cl) \qquad \rightarrow \qquad [-\,CH_2 - CH\,(Cl)-\,]_n$$
$$\text{polymerize}$$

The PVC is a thermal conductor meaning it absorbs or gives off heat very easily. Hence, vinyl seats get warmer quickly in the hot sun and colder in the wintertime.

Now she turns the ignition on and within awhile, computer voice activated module warns her about 'low cooling fluid' suggesting that the anti-freeze in the cooling system is low. She turns the ignition off, opens the hood and pours some anti-freeze into the reservoir until it reaches the maximum level indicator.

Chemistry in Service

Anti-freeze: Thermochemistry is the study of heat exchange associated with a substance or a chemical reaction that is also responsible, in conjunction with solution chemistry, for the development and use of antifreeze in automobiles. Anti-freeze is the homogeneous mixture of water (H_2O) and one or two organic solvents, namely, ethylene glycol ($OH-CH_2-CH_2-OH$), propylene glycol (CH_3-CH $(OH)-CH_2OH$). It is used to prevent the engine block from melting down at high temperature and freezing and eventual cracking at low temperature. As the circulating fan mounted on the engine block quickly cools down the hot antifreeze coming from the engine, it is quickly circulated through the engine again by the pump. This process takes place in a continuous manner. Ethylene glycol is used instead of 100 % H_2O because there is a great difference in their specific heat capacities and melting and boiling points (melting point is the temperature at which solid melts into liquid and boiling point is the temperature at which the liquid boils into gas). As we know water has a freezing point of 32^0 F (0^0 C) and a boiling point of 212^0 F (100^0 C). Antifreeze and water in the ratio of 1:1 has a freezing point of -22^0 F (-30^0 C) and boiling point of 227^0 F (108^0 C). Due to this, it doesn't evaporate or boil away as water does at the engine temperature or does not freeze as the water does in cold climate conditions. The lowering in the freezing point below the freezing point of pure solvent (in this case water) is known as the *freezing point depression*, and raising the boiling point above the boiling point of pure solvent is known as the *boiling point elevation*. Therefore, the anti-freeze has the dual purpose - keep the engine block from freezing in the wintertime and from over heating and eventual melting down of the engine block in the summer time.

She closes the hood, gets back into the car, turns the ignition on again hoping no other problems surface and starts to drive off the car. She programs the car computer for her destination. As a good driver as she is, she always turns her eyes all around including the dashboard like a radar to detect any unforeseen events. To her surprise, computer voice again warns her about 'near empty tank' suggesting that the amount of gas in the gas tank, according to its mileage/gallon ratio calculation, is not enough for her to drive to school and drive back from school. Mercury asks the computer to locate the nearest gas station with lowest price. No fear! Says the computer. The gas station is just around the corner. So, she drives to that gas station and parks the car in a designated pump area. Sitting in her driver's seat, she instructs the robotic arm on the pump to fill up her tank with regular (87% octane) unleaded gas (please note that all new cars in USA are equipped with catalytic converter and hence only the unleaded gasoline of at least 87% octane rating must be used). By the way, the gasoline is called 'gas' in the United States while it is known as "petrol" in most other countries.

Chemistry in Service

Gasoline: Gasoline in general is just very refined petroleum. It is a mixture of hydrocarbons (organic molecules containing hydrogen and carbon only) containing C_5 to C_{12} atoms and when burned they produce the carbon dioxide and water. This process is called the combustion (combustion is a rapid process of combining the substance with oxygen). For example octane, C_8H_{18}, the combustion of which in presence of plenty of oxygen takes place according to,

$$2C_8H_{18}(g) + 25O_2(g) \quad \rightarrow \quad 16\,CO_2(g) + 18\,H_2O(g)$$

Various grades of gasoline are sold at the pump. These grades are expressed, for the sake of consumer easy understanding, in terms of octane rating. The octane rating is a numerical expression of a particular gasoline's ability to resist an engine's knocking or pinging. The grades typically vary from 87 to 93.5. Higher the octane rating means higher the rate of smooth burning meaning higher the efficiency of the car. [**Octane rating**: Burning properties of hydrocarbons are structural dependent. The branched- chain hydrocarbons burn smoothly while the straight-chain hydrocarbons burn too rapidly making a sound known as 'knocking' or 'pinging.' This not only reduces engine power but also may cause damage to the engine. Isooctane (2,3,4-trimethylpentane) is selected as the standard in assigning octane ratings because of its smooth burning. Octane ratings seen on gas-pumps are arbitrary numbers used in rating the relative knocking property of gasoline. Heptane (C_6H_{14}) – a straight-chain hydrocarbon burns very rapidly and hence it is assigned the octane rating of 0, while isooctane burns smoothly and hence it is assigned octane rating of 100. In determining the octane rating, first the gasoline is burned in a standard test engine and its knocking property is recorded. Then the knocking property of mixture of heptane and isooctane is adjusted to match the gasoline property. For example, 87% octane means that gasoline has the same knocking property as the mixture of 13% heptane and 87% isooctane.] There are certain drawbacks driving the car that one should be aware of. In addition to production of carbon dioxide and water, burning of gasoline also produces some undesirable substances like carbon monoxide (CO), unburned hydrocarbons, and nitrogen oxides (NO_x). Carbon dioxide is a colorless gas exhaled by animals is harmless and needed for the photosynthesis process. It is the major constituent of car exhaust and is often implicated as being the major gas responsible for the perceived global warming trends in recent years. Carbon monoxide is also colorless and odorless but a very poisonous gas. It too is found to be a major constituent of car exhaust, especially at engine idle. As little as .3% by volume can be lethal within 30 minutes. In order to eliminate undesired substances from the car exhaust, the petroleum industries are trying to find a more environmentally safer source of energy for cars. Recently they have introduced a product known as oxygenated gas. Oxygenated gas is ordinary gas with the addition of ethanol or ethyl *tert*-butyl ether (ETBE), derived from isobutylene and ethanol, and methyl *tert*-butyl ether (MTBE), chiefly derived from isobutylene and methane, which provide oxygen. Among these, ethanol offers most oxygen.

$$C_2H_5OH \qquad C_2H_5 - O - C(CH_3)_2 - CH_3 \qquad CH_3 - O - C(CH_3)_2 - CH_3$$

$$\text{Ethanol} \qquad\qquad \text{ETBE} \qquad\qquad\qquad \text{MTBE}$$

Clean Air Act amendments of 1990 mandates those areas that do not meet federal carbon monoxide standards must use oxygenated gasoline during winter months when carbon monoxide levels in the air are the highest. Gasoline in these locations must containing at least 2.7% oxygen by weight and must be used from November through February, a span of four months during the winter. The 2.7% oxygen requires 15% by volume of MTBE in gasoline. Other comparisons are listed below.

	Ethanol	ETBE	MTBE	Summer RFG[#] Southern grade
Octane rating	115	112	110	88
Blending RVP (Reid vapor pressure)	18	4	8	7.2 maximum
Oxygen content (weight %)	35	15.7	18.2	2.0 maximum
Oxygenate % by volume to achieve 2.0 weight % oxygen	5.7	12.8	11	

#Source: G. Peaff, C & E News, September 26, 1994.

There are of course few drawbacks in using the oxygenated gasoline. First of all, the gasoline burns rapidly driving up the cost of driving the car. These added chemicals have been suspected in causing various illnesses like nausea, headache, dizziness, and eye irritation.

After having taken the gasoline and paid the gas station attendant through her internet bank account by accessing it through her car computer, Mercury rushes to school, parks the car in the student's parking lot and runs to the chemistry building to attend the chemistry lecture. Her first morning class in chemistry deals with elements and their significance. She is very attentive in the classroom and takes careful and accurate notes on the subject matter as professor writes on the blackboard with chalk. The professor starts with the sentence, what are elements?

Chemistry in Service

Elements: Elements are everywhere in nature, which are the building blocks of our nature. Elements, as we know them today, differ from man's first thoughts (See Appendix E. A Brief History of Chemistry, for more discussion on early thoughts on elements). Greek philosophers of the fifth century B.C. believed that there are only four basic elements, air, earth, fire and water - from which everything was made in different proportions. However, the Greek philosopher Democritus believed that all matter is composed of very tiny, indivisible particles - he labeled them as *atomos* (Greek word for indivisible). Even though other well-known Greek philosophers like Plato and Aristotle did not accept Democritus's concept, his concept lingered around for centuries until 1808 when an English schoolteacher and a scientist (John Dalton) formulated the precise definition of elements using Democritus's concept. Now we know that all elements are made up of atoms. Most of the elements known on Earth, whether natural or man-made, are also known to be components of stars and other planets.

Once elements are identified, chemists classify them, and organize them according to their diverse properties. Elements have been charted in the form of a periodic table. D. Mendeleev, a Russian Chemist developed one of the original periodic tables. He was the first one to arrange systematically according to their atomic weights. The most fascinating account of Mendeleev's table was that he left empty spaces in the periodic table for yet to be discovered elements and using his table, he was able predict the properties of these yet to be discovered elements. The currently accepted periodic table is similar, but assigns each element a number that distinguishes it from any other element. This number equals the number of protons in the nucleus of that element known as the atomic number. The atomic number also determines the number of electrons needed to balance the positive charge of the nucleus. The most basic division of the elements in the periodic table is into metals, nonmetals, metalloids, and noble gases.

Elements in the periodic table are listed horizontally in order of their increasing atomic number as well as atomic mass, forming a series of rows called periods. When this is done the elements falling in the same vertical column known as groups exhibit similar chemical properties. In addition to this main periodic table, two rows of elements, lanthanide series and actinide series are placed below the main table. Lanthanide series elements exhibit similar properties as lanthanide but due to lack of space to accommodate them next to lanthanide, they are placed under the periodic table and labeled as ' lanthanide series' meaning the elements starting with lanthanide. Same explanation also holds for actinide series.

Mercury, as she listens to the lecture, suddenly becomes aware of the situation that a student sitting next her has a cast on her leg, which looks very similar to the chalk the professor is using to write on the blackboard. Now she starts to ponder herself on the similarity between the chalk and the cast.

Chemistry in Service

Chalk: Professor explains about elements and the periodic table by writing on the blackboard with a chalk. The chalkboard itself is a smooth and dark board usually made from slate, glass, or wood. The chalk used on the blackboard consists primarily of calcium carbonate ($CaCO_3$). The other common names for calcium carbonate are marble and lime stone. It is fine-grained white limestone, which can be used to write.

Cast (Plaster of Paris): The cast used to set a broken arm or leg is nothing but calcium sulfate dihydrate ($CaSO_4 .2 H_2O$). Like chalk, it is also another salt of calcium. Plaster of Paris, on the other hand, is a calcium sulfate monohydrate [$(CaSO_4)_2 . H_2O$ (s)]. To prepare the plaster of Paris, calcium sulfate dihydrate is heated to lose 3 moles of water according to the following reaction.

$$2 CaSO_4 . 2 H_2O \text{ (s)} \rightarrow (CaSO_4)_2 . H_2O \text{ (s)} + 3 H_2O \text{ (l)}$$
calcium sulfate dihydrate(cast)   plaster of Paris

This is a reversible reaction and the cast is formed when water is added to plaster of Paris.

Mercury keeps on continuing to take notes without comprehending that she is recording every thing professor says on sheets of paper with a pencil and ink, which in fact are brought to life by the magic touch of chemistry.

Chemistry in Service

Paper and Pencil / Ink: The paper is composed of layered fibers. It is porous, which allows the ink from pens to penetrate its structure. There are various kinds of inks, for examples, the red ink is made from a red ore of mercury. Ink can also contain powdered carbon.

To understand how the pencil's lead adheres to the paper it is necessary to understand the elements in the pencil. The lead in the pencil is composed of graphite mixed with clay (clay allows the graphite to retain a harder consistency). Graphite is an allotropic form (different molecular or crystalline form) of carbon, which is a soft, dark black solid with metallic luster. In addition, the natural carbon also has two allotropic forms know as diamond and Buckminsterfullerene, commonly refers to as *buckyball*.

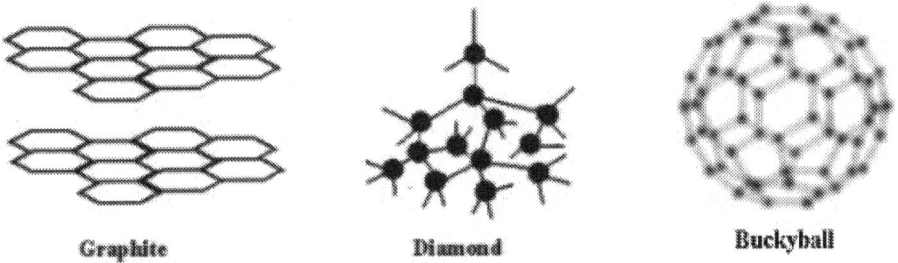

Graphite   Diamond   Buckyball

The carbon atoms in graphite are covalently bonded to one another in hexagonal fashion (benzene ring) and these hexagonals are fused with one another forming layers or sheets. The sheets are held together with the weak overlapping of pi-orbitals, which is responsible for their slipperiness. Many properties of graphite are anisotropic. Commercially, the graphite is produced from either charcoal or coke. There is no strong bonding between the layers in graphite leading towards an empty space. This layering is very useful because they tend to split apart so that graphite can flake easily, therefore allowing the flakes to mark the paper if rubbed across it. The marks are the flakes rubbed off and left as the graphite moves.

The carbon atoms in diamond are tetrahedrally bonded to one another. It is the hardest known material known to mankind, and hence it is used commercially as cutting and shaping agent. It is the most expensive natural carbon.

The buckyball consists of 60 carbon atoms arranged in six-membered and five- membered rings, which are fused together in the shape of a soccer ball. This allotropic form has been discovered in the later part of the last century. For greater details about this form Appendices B and C may be consulted.

After finishing the lecture, she starts walking towards the chemistry laboratory. As she is strolling along, she begins to wonder about the relationship between elements and life.

Chemistry in Service

What is life? Elements are all around us - they are basic ingredients of life and it is how they combine which makes the difference in what they form. Inanimate and animate objects both are composed of elements. Of these, what can we classify as living organisms? Living organisms are considered just as machines according to the mechanistic point of view, and hence no sharp contrast is drawn between living and nonliving systems. Then question comes, what is life? Life has been defined to have the properties of the ability to reproduce, respond to external stimuli, and to grow. Inanimate systems share these properties with animate systems.

An attempt to recreate living material as it appears on Earth currently by experiment proved that amino acids could be created (the building blocks of proteins that are essential components of living things on Earth) by passing an electric shock through a mixture of the primary components of the Earth's atmosphere (oxygen and nitrogen). All Earth life is composed of mainly carbon, hydrogen and oxygen with smaller quantities of other elements, and hence the life on Earth is labeled as **carbon based.** The mechanism by which all lives work is DNA (Deoxyribonucleic acid). DNA, based on carbon chemistry, might be able to be re-arranged to create other life forms. Other elements might originate life forms such as silicon, which is the primary component of computers, and hence computers are labeled as a life of silicon **based.**

Computer-generated artificial life can also reproduce, evolve into something other than its original form, therefore these creations can be classified also as **alive.** Scientists hope that in the future computer-generated artificial life can co-exist with organic life.

May be life exists somewhere else in the universe, composed of elements foreign to the Earth's composition as we know it. As an example, Mars' atmosphere is cold with thin air; elements capable of sustaining life under these conditions may exist. There may exist another life form composed of elements unique to that system - Jupiter or Saturn might already now be encouraging an alternate-Earth chemical life.

May be another form of life stranger to the organic life may very well coexists on this planet that we do not even know about it, due may be stringent requirements imposed on the organic life.

If an alternate life exists in this universe, how can we acknowledge if it is not in any physical form? Human eyes can see only the physical objects; humans can feel and perceive only certain things. Instruments, on the other hand are created to see and recognize the things that humans want them to see or recognize, go beyond human's ability. But the other life might be beyond their reach also. Hence, it appears that different kinds of instruments are needed to recognize other life, but how can it be possible? We even do not know what is the other life is!

Mercury now goes into the chemistry laboratory to do her chemistry experiment. Today's assignment involves an understanding of the SI (System of International) units and determination of the density of a liquid and a solid. Even though, the density is a simple easily determinable physical property, it remarkably yields lots of information about the nature.

Chemistry in Service

SI Units: Science is largely based on mathematics and logic. Because scientists need to convey information globally, a standard mathematical system is inevitable. Thus, the international system of units (SI units) is developed. This mathematical system draws from the metric system, a rationally decimal-based system of units for length (meter), mass (grams) and volume (liter). Each of the base units may easily be converted into another unit by means of multiplication or division or moving of a decimal point. The English system of equivalents include: 1 yard = .9144 meter; 1 quart = .9464 L; 1 pound = 453.6 grams. Even more, the metric system demonstrates an adjoinment of length and volume (as 1 ml = cm^3) and volume and weight (as 1 gram = 1 ml of water at 4^0 C). The essential SI units utilized in the laboratory are meter (m) for length, kilogram (1 kg = 1000 grams) for weight, second (s) for time, Kelvin (K) for temperature, ampere (amp, electric current), mole (mol, amount of a substance), candela (cd, luminous intensity).

Density: Density is a physical property of a substance (physical properties are defined as the properties change in physical appearance but not change in chemical composition) and is defined as the ratio of the mass (weight) of a substance to its volume. In essence, the density is the mass per unit volume, i.e. density = mass / volume. Most commonly, density is given in units of grams/milliliter (g/ml) or grams/cubic centimeter (g/cc). It should be noted, however, that in reality mass differs from weight. - mass is a constant amount of matter in an object while weight is a measure of the force of attraction of the earth's gravity on an object.

Methanol (CH_3OH), a liquid and copper (Cu), a solid, are given to determine the densities. Mercury first determines the density of methanol by the following procedure; she first weighs a clean and dry 250 ml empty pycnometer and then fills it with methanol and reweighs the pycnometer. She calculates the mass of methanol by taking the difference between second and first weights. To determine the volume of the pycnometer accurately, she fills it with water, wipes out all the adhered water on outer surface of the pycnometer with paper towel and weighs it. Difference between this weight and the weight of empty pycnometer gives the mass of the water required to fill the pycnometer. She converts this mass to volume by using the density of water 0.9998 at 25^0 C. Then she divides the mass by the volume to yield 0.79 g/ml as the density of methanol.

To determine the density of a solid copper, first Mercury weighs an empty beaker and then adds the copper sample to the beaker and reweighs the beaker. By subtracting the mass of the empty beaker from the mass of beaker and copper sample, she gets the mass of the copper sample. The volume of solids, in general, is not determined as easily as liquids. If they are regular solids, such as, cube, cylinder, all required dimensions are measured and the volumes are calculated using appropriate mathematical formulas. However, if they are irregular solids, such as, chunks or small shots, as is the case with the copper sample, the volume is obtained by the displacement technique. To get the volume of the copper, Mercury first adds 30 ml of water to a graduated cylinder and then adds the copper sample to the graduated cylinder. The difference or increase in volume gives her the volume of the copper sample. She determines the copper density by dividing the recorded weight of the copper sample by the volume. The result yields 8.9 g/ml as the density of copper. When Mercury compares the experimental densities of methanol and copper with the literature values, she realizes that her values are closer to the literature values verifying that density is in fact a fixed constant ratio.

At this point, Mercury began to wonder about the kind of balance she is using to weigh the objects. The current balances give the mass of any object in grams. However, Mercury knows for the fact all substances are made up of molecules and atoms. Is it not wonderful, she said to herself, if the balances also show the equivalent number of entities like molecules or atoms in addition to grams? Well, this issue has been addressed in Appendix D.

Significance of density: From this experiment it is seen that the density of the copper is higher than the methanol. In general, all the solids in nature have higher densities than liquids, which in turn have higher densities than gases. Such difference in the densities can be explained in terms of packing of atoms in each physical state of matter - the atoms are very tightly packed in solids due to their much stronger interatomic attractive forces, are less tightly packed in liquids due to their less stronger interatomic attractive forces, and are far apart from each other in gases due to much weaker interatomic attractive forces or much stronger interatomic repulsive forces. Temperature and pressure affect density. Although mass remains constant, changes in temperature and pressure alter volume. Temperature change only minimally affects the density of solids and liquids; however, it greatly influences gas densities as seen by the fact that hot air is less dense than cold air. A gas expands when heated, thereby increasing volume and subsequently decreasing its density. Even though, the density is a simple physical quantity, it is very useful. For instance, in determining the charge on an automotive battery and in determining the sugar concentration for fermentation purpose in wine making.

After having attended a chemistry lecture and finishing her labora-tory assignment, Mercury is starving for lunch. She and her few friends rush towards a cafeteria to replenish their lost energy. Mercury, being a very health conscious person, picks up one low fat yogurt (150 calories per 6 oz low fat) container, a bowl of salad consisting of lattice (3 small leaves 3 cal), carrot (one medium 25 cal), tomato (raw, one medium 30 cal), spinach (1/2 cup, 20 cal), and squash (1/2 cup 25 cal), and a small amount of salad dressing.

Chemistry in Service

Yogurt: Yogurt is bacteria fermented product of milk. The fermentation process depends upon the production of lactic acid (CH_3-CHOH-COOH) by a mixed culture of the bacteria *Lactobacillus* and *Streptococcus thermophilus*. When milk comes in contact with bacteria, the milk sugar lactose is broken down by the bacteria for its energy while producing lactic acid. The acid causes the fatty droplets to coalesce and curdling the milk - yogurt. The word **yogurt** is in usage in USA and the word **curd** is in practice in Europe and Asia.

Salad: Salad is mainly consists of various slices of vegetables, which contains numerous proteins, minerals, and various organic and inorganic molecules.

Salad dressing: Salad dressing is an emulsion of one kind or the other, may look like either a homogeneous or a heterogeneous mixture of various chemicals of different physico-chemical nature. A typical salad dressing may contain natural ingredients like water, corn syrup, sugar, vinegar, tomato paste, copped pickles, salt (NaCL), dried onions, lemon juice, and chemical like cellulose gel, xantham gum, potassium sorbate and calcium disodium EDTA, propylene glycol alginate, phosphoric acid(H_3PO_4), artificial flavors, and artificial colors.

Serving size of 30 ml contains about 130 cal. Sodium content is high usually about 240 mg. Artificial flavors and colors have been known to cause cancer in laboratory animals.

After lunch, Mercury attends another lecture for the day, this time in biology, dealing with the photosynthesis, a process that is vital for the survival of all plants in nature. She tries her level best to digest the chemistry of photosynthesis. Photosynthesis is the light-induced process that has been taking place in the nature since the birth of green plants. Of course, the required condition is the presence of solar energy.

Chemistry in Service

Photosynthesis: The purpose of photosynthesis is to produce food (glucose and other simple sugars) from carbon dioxide from the air and the water from the soil by using light energy from the sun. This is a complex process that is summarized by the following reaction.

$$6\ CO_2(g) + 12\ H_2O(l) + \text{light energy} \rightarrow C_6H_{12}O_6(s) + 6\ H_2O(l) + 6\ O_2(g)$$

From the above equation it is evident that it requires 6 moles of carbon dioxide (264 g) and 12 moles of water (216 g = 217 ml) to synthesize one mole of glucose (180 g) or 620.5 g of carbon dioxide and 545 g (= 545 ml) of water to produce one pound of sugar.

Photosynthesis proceeds in two stages that are known as the light reactions (light-dependent) and dark reactions (light-independent), and each stage has its own set of reactions. In the light reactions of the photosynthesis, solar energy is converted to chemical energy where solar energy is absorbed by the chlorophyll that drives a transfer of electrons from water to an acceptor $NADP^+$, which temporarily stored the electrons. The water from the soil is split in this reaction giving off O_2. In addition, the light reaction also makes use of solar energy to reduce $NADP^+$ to NADPH(Nicotinamide adenine dinucleotide) by adding a pair of electrons along with a proton. The light reaction also generates ATP (Adenosine Triphosphate) by adding inorganic phosphate (P_i)

group to ADP (Adenosine Diphosphate), a process is known as pho-tophosphorylation. The reaction for this stage is,

$$12\ H_2O(l) + 12\ NADP^+ + 24\ P_i + 24\ ADP \rightarrow 12\ NADPH + 12\ H^+ + 24\ ATP + 2\ H_2O(l) + 6\ O_2(g)$$

light

Thus, solar energy is transformed into chemical energy in the form of two substances, ATP, primary carrier of chemical energy in the cell, and NADPH, a source of energized electrons.

In dark reactions of the photosynthesis, the sugar is produced. The CO_2 from the air is incorporated into organic material, a process labeled as carbon fixation. The fixed carbon is then reduced by NADHP to sugar by addition of electrons. In addition, the dark reactions also use the energy from ATP. The chemical reaction for this process is,

$$6\ CO_2(g) + 24\ ATP + 12\ NADPH + 12\ H^+ \rightarrow C_6H_{12}O_6\ (s) + 6\ H_2O(l) + 24\ P_i + 24\ ADP + 12\ NADP^+$$

How these two stages work together is shown below:

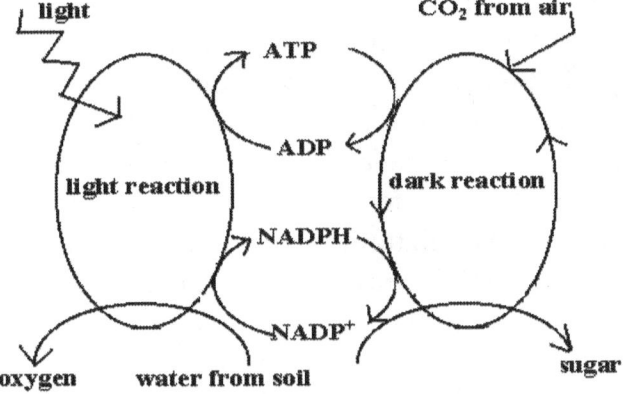

Mercury is not only interested in science but also in art, as one day she wants to be a different blend of generation by combining her knowledge of science, art and literature. To stimulate her desire, she is also taking an art course. Her next class is the art class dealing with painting.

Chemistry in Service

Painting: Even in the art class various chemical forces are at work: there are solutions, emulsions, mixtures, water-based and oil-based substances, and these only represent a fraction of chemistry's involvement with art. Mercury being a dedicated painter, she constantly interacts with materials whose origins and properties are best explained through chemistry. For instance, when an artist creates a painting, there is a great deal of chemistry involved: materials (charcoal, crayon, pastel, oil paint, acrylic paint), supports (the materials the painting actually is put onto, such as canvas/linen, the ground, such as gesso, and the medium, which is used to thin or thicken the paint texture, such as linseed oil/poppy seed oil, turpentine), and finishes (such as damar varnish, or resins), all feature a variety of chemical reactions and properties that are interesting for the artist to understand for both artistic means and safety precautions. Recently, the importance of physicochemical reactions as they affect the physical behavior and character of liquid paints has been addressed by the makers of industrial paints, and with an increasing demand, a need has arisen for pigments treated to meet specifications that are desirable for both the use and safety of artists. Oils and resins are variable mixtures of highly complex organic chemical compounds, among which, are acids or compounds of an acidic nature that are capable of reacting chemically with alkalis under proper conditions, for instance, the linseed oil that is preferred by many artists as a medium for oil paints should have a pH less than 6.5. The most abundant constituent of linseed oil is believed to be dilinoleo-linolenin.

As an oil paint dries off, it undergoes changes that are the result of physico-chemical reactions taking place between pigments and the oil, as well as by changes brought about by the oxidation of the oil. These are of course complex reactions vary from pigment to pigment. The quality of the oil paint, although depends on the nature of the pigment, various other factors like consistency, drying speed, extent of oxidation, flexibility, hardness, durability, and color stability are also deciding factors.

Turpentine, which is almost a pure and refined chemical (it contains 92-96% of pinene) is often used to clean the artist's brushes and palette, and can also be used as paint thinner. Turpentine is a product of pine distillation, and has the property of polymerizing as it ages and passes into other heavier, more viscous stages that are less volatile and usually have a less agreeable odor. Sunlight, air and moisture cause turpentine to be rapidly oxidized, the reason why it is best to use fresh material and store it in full containers, brown bottles or tin cans.

For the artist, a prime consideration is choosing colors. The advancements of chemistry in the last few centuries have benefited the world of fine arts; the synthetic manufacture of painting supplies has brought about not only more easily accessible supplies, but also even new colors that were unknown to earlier artists. For instance, two German chemists, C. Graebe and C.Liberman in 1868, first synthesized alizarin crimson. Alizarin, a quinone is the coloring principle of the madder root (an herb), which is found in France and earlier was known in ancient Egypt, Persia and India.

Alizarin

This was an important discovery in the history of organic chemistry, as alizarin was the first of the natural dyestuffs to be made synthetically. Alizarin crimson is made with aluminum hydroxide (Al (OH) $_3$).

Different colors often have different properties of intensity, luminosity, and deterioration. For instance, white lead ($PbCO_3$) and zinc white (ZnO) may appear to be the same color, but the two employ different properties for use by the artist. White lead is the most important of all the lead pigments, which was used as a pigment in ancient times. Pure white lead in oil is factored as an outside white paint because it does not crack and leaves a satisfactory surface for repainting, which is one of the main advantages of painting with oil paints. It is darkened, sometimes, by contact with sulfide pigments and hydrogen sulfide (H_2S) in the air because of the formation of black lead sulfide (PbS), and is readily soluble in dilute mineral acids and in acetic acid with effervescence (CO_2). Zinc white, or zinc oxide (ZnO), on the other hand, is lighter and more bulky than white lead, and requires more oil to form a paste than white lead. It's tendency to become brittle and crack is considered a disadvantage, but it is not affected by bright sunlight. Knowing the properties and production procedures of varying pigments can help the artist choose the appropriate paint for his/her work. There are instances when a light, transparent white film is needed to achieve a desired effect in a painting while in other instances, heavy, thick paint is more appropriate. Knowing what paint is more applicable to the surface and subject of the painting is a great advantage to the artist. Another example of chemistry related to color is viridian. Viridian is the transparent, bright green, hydrous oxide of chromium (Cr_2O_3. $2H_2O$). The above example of zinc white, lead white, and viridian show that the pigments comprise a wide variety of chemical properties. Although there are a few that are complex metallo-organic compounds, such as Prussian Blue and emerald green, most are inorganic coloring materials such as oxides, sulfides, carbonates, chromates, sulfates, phosphates, and silicates of the heavy metals. Other chemicals widely used for pigments are: Yellow ochre (Fe_2O_3

.H_2O) , raw umber ($Fe_2O_3+MnO_2+H_2O$), cadmium red (CdS (Se)), emerald green ($Cu(C_2H_3O_2)$ $3Cu(AsO_2)_2$), and charcoal (C and bone black) ($C+Ca_3(PO_4)_2$). It should be noted that a material has color because of it has selective absorption for the colors of white light.

Some paints and pigments are toxic. There are two types of toxic materials related to the art, the first type including those substances that are potentially toxic but can be handled safely, and the second type being those materials that are of such toxicity that they should be altogether avoided. Flake white has been manufactured for more than 2000 years and for about 500 years was the only opaque white oil color available to artists. But by taking several precautions, such as, using it only in oil-paint form and never handling the dry pigment, an artist should be able to use it without any endangerment. Similar precautions, including thoroughly washing hands before touching food, should be followed when using a color called naples yellow, which is the only other approved lead pigment. The more toxic paints are emerald green, and a grade of cobalt violet, which contains arsenic, and hence their use should be avoided.

In general, there are a whole host of dangers that may inhibit artist's ability, such as fatigue, headaches, anemia, damage of kidneys, and paralysis. It is thought that the most hazardous materials encountered by painters are solvents, of which even the least harmful of them could cause physical disorders if improperly handled. Because of their low volatility, some like turpentine and mineral spirits are relatively safe for studio use, but acetone (dimethyl ketone (CH_3COCH_3)), ethyl alcohol (C_2H_5OH), and benzene (C_6H_6) can cause great damage, from an irritation to lungs by acetone, to damage to the nervous system and blood-forming organs by benzene, which simply should not be used.

After having done some excellent paintings, Mercury feels that she needs to be refreshed before she can move on to the library for some reference work for her classes. So, she walks towards the snack bar, which is just around the corner, and grabs a can of cola soft drink and a Snickers candy bar, so that she can temporarily relieve her hunger and thirst. Momentarily she starts chewing on the candy bar while sipping the soft drink and at the same time propelling towards the library. By the time she arrives at the library, she finishes her candy bar and soda. Mercury is no 'litter bug' and always tries to maintain the clean environment and the beauty of the campus. Therefore, she deposits the soda can in a container that is designated as 'recycle cans', and throws away the candy-wrap in the wastebasket placed in front of the library.

Chemistry in Service

Cola drink and Candy bar: The cola drink contains a large amount of sugar in the form of corn syrup. In fact, drinking one can of cola is equivalent to eating nine teaspoons of pure granular sugar, which is largely responsible for the caloric content of 155 calories per 12-ounce can. Cola also contains 39 grams of carbohydrates, as well as about 100 milligrams of caffeine, which is known to have several negative effects on humans. First of all, it is psychologically addicting and secondly, it is a vasoconstrictor meaning it makes blood vessels constrict. Studies have shown that it may enhance chromosomal damage brought on by certain carcinogens. All these are very good reasons to avoid caffeine, but on the other hand, caffeine is a stimulant, which can be helpful when being groggy.

Snickers bar has an incredible 510 calories. It looks like Mercury is going to have to work extra hard during her workout to burn off all these calories. In addition to the high caloric content, the candy bar also contains 64 grams of carbohydrates, 9 grams of protein, 280 milligrams of sodium, and a high 24 grams of fat. All an all, a candy bar is probably not the healthiest of snacks available, but for students, it is quick bite on.

Now Mercury is ready to do some reference work for the classes she attended earlier. Although the library seems to be a very sedentary place, active chemical processes are constantly in effect. Paper materials, because of their composition, tend to be acted upon by outside forces. The library is a storehouse for information, and hence the ability to produce and reproduce that knowledge is very important. This is why the use of computers and office supplies, such as copy machines, have had such a positive effect in the dissemination of information/knowledge.

Chemistry in Service

Paper: Paper is a very general name for the substance used for variety of purposes - writing or printing upon or packaging, wrapping, etc. The name paper is derived from the *papyrus* plant. Similar material was first produced by the ancient Egyptians. Paper is manufactured from a mixture of various fibers, mainly of vegetable origin, which are mixed with a large quantity of water and shredded to very fine stage. The mixture is then treated with size (a glue-type admixture which makes it water resistant) and filler, such as clay or chalk to give some special property to the paper. After this, it is poured on the wire screens on which the water is extracted.

The fibrous material used in modern-day papermaking is wood from deciduous tress (beech, poplar, birch, chestnut, and eucalyptus), coniferous trees (fir, spruce, and pine), grasses (wheat, rice, barley, oats, straw from rye), alpha grass, cotton, and bamboo.

It has been known that pure cellulose fiber is the most resistant and permanent part of paper affected only by very high temperatures, acids, alkalis, and strong bleaching solutions (mainly oxidizing agents). People are the "worst enemy of paper," followed by other elements such as air, light, darkness, heat, moisture, vermin, acid and fungi. Librarians, archivists, and conservators generally store historically and artistically important documents in such a way, including using acid-free storage containers, as to prevent the ill effect of these factors. Laboratory tests have shown that if paper could be stored in a dry, close to freezing atmosphere, it could last almost indefinitely. However, under those conditions, people could not read or work with the books and papers. A compromise is generally carried out whereby the rooms are kept at about 68^{0}F (20^0 C). Air-conditioning seems to help libraries to rid off some types of fungi and acid. Ultraviolet rays, which are present in natural and artificial lighting, are quite harmful, causing ink to fade and the destruction of paper. But by keeping direct sunlight out of processing, reading and exhibit areas, and also by using protective sleeves for fluorescent light fixtures, these problems may also be alleviated. But it is human contact with paper that appears to pose biggest problem because the oils, dirt and detergents present on human hands and also some chemicals exhaled by breathing the same environment tend to react with paper.

There is presently experimental work being done that is attempting to perfect a de-acidification process, which can be applied to the leaves of books that do not require lamination or rebinding. In the process of de-acidification, several steps are followed: the paper is treated with a mild alkali which initially serves to neutralize any acid that is present and is then converted into a compound that remains in the fibers of the paper to act as a buffer to neutralize any further acidity that may develop.

There are some solutions that are dangerous to paper materials such as hydrochloric acid (HCL) that is used for bleaching of fibers. If it is insufficiently removed through washing processes, residual hydrochloric acid can exist in paper. Because it is the corrosive acid, and it's corrosive effect is particularly deleterious to the cellulose fiber as it causes the cellulose chain to break into even smaller units, and possibly resulting complete hydrolysis of the paper.

*After doing an intensive reference work, Mercury feels exhausted men-
tally, and she needs to relax a little bit before she goes back home for the
day. Only way she relaxes is to do some workout in the gymnasium. So,
she walks to the gymnasium, which is about few blocks from the library.
She spends about an hour in gymnasium doing some workout including
swimming. She realizes that any kind of exercise program depends on the
training effect for its results. This merely means that when the body is
subjected to unusual stress over a period of time, it adopts itself so that it
can deal more effectively with that stress (sounds familiar! Le Chatelier's
Principle). But living and exercise is work, and work requires energy
(thermodynamic principle). The production and release of this energy is
nothing more than a complex set of chemical reactions.*

Chemistry in Service

Exercise and ATP: ATP (Adenosine triphosphate) contains adenosine
nucleus, ribose nucleus, and three phosphate groups through two phosphoric anhydride bonds, and considerable energy is released when ATP
is hydrolyzed to

ADP(Adenosine diphosphate) and further to AMP (Adenosine
monophosphate). Energy is stored in phosphoric anhydride bonds.

Hydrolysis is exergonic process releasing 7.3 kcal for every mole of ATP that loses the terminal phosphate group.

ATP $\rightarrow$ ADP + P + Energy

The cells also store glycogen but most of the energy is in substances called phosphages. Creatine phosphate, the phosphagen of vertebrates, has a high-energy phosphate bond that it transfers to ADP to make ATP. After food is digested and distributed to cells, sugar and other organic molecules are consumed as fuel in the process of cellular respiration. With the help of oxygen, cellular respiration harnesses the energy stored in the fuel molecules for work, such as contraction of muscle cells.

The main engine of the eukaryotic cell is the mitochondrion, where most of the enzymes and other metabolic gears of respiration are located. The main fuel for respiration is the sugar glucose, and the exhaust is carbon dioxide and water. The overall process can be summarized as follows.

$C_6H_{12}O_6$(glucose) + $6O_2$ (oxygen) $\rightarrow$ $6 CO_2$(carbon dioxide) + $6 H_2O$(water) + energy

While one exercises, one spends tremendous amounts of energy in the form of ATP. To keep on working, the cell must regenerate its supply of ATP from ADP and inorganic phosphate, the process is known as phosphorylation of ADP, which is endergonic as hydrolysis is exergonic. The main purpose of respiration is to provide energy for ATP synthesis.

Animals require a continuous supply of oxygen (O_2) for cellular respiration, and they must expel carbon dioxide (CO_2), the waste product of this process. It is important to understand the difference between the gas exchange- the exchange of oxygen and carbon dioxide between the animal and its environment, and the metabolic process of cellular respiration- supply of oxygen and removal of carbon dioxide. Respiration is a cumulative function of three metabolic stages; (a) glycolysis, (b) the

Krebs cycle, and (3) the electron transport chain and oxidative phosphorylation.

Glycolysis and the Krebs cycle are catabolic pathways that decompose glucose and other organic fuels. Glycolysis begins the degradation by breaking glucose into pyruvic acid (CH_3-CO-COOH). The Krebs cycle, located within the mitochondrion, completes the job by decomposing a derivative of pyruvic acid to carbon dioxide. A limited amount of ATP synthesis is coupled directly to specific steps in glycolysis and Krebs cycle. However, these two stages function mainly to supply energized electrons to drive oxidative phosphorylation, which accounts for most of the ATP made during respiration. The metabolic machinery for oxidative phosphorylation includes an electron transport chain, a group of molecules built into the inner membrane of the mitochondrion. Electrons removed from food molecules during glycolysis and the Krebs cycle are pulled down the electron transport chain to a lower state of energy by oxygen, which is very electronegative.

The electron transport chain makes no ATP directly at specific points along the electron transport chain, protons are pumped from the matrix to the inter membrane space. The electron transport chain thus functions as an energy converter, using the fall of electrons from food molecules to oxygen to store energy in the form of a proton (H^+) gradient across the inner member of the mitochondrion. It is the potential energy stored in the proton gradient that function as a mill to harness the exergonic passage of H^+ to drive the endergonic phosphorylation of ADP.

Human muscle cells make ATP by lactic acid fermentation when oxygen is scarce. This occurs during the early stages of strenuous exercise, when sugar catabolism for ATP production out paces the muscle's supply of oxygen from the blood. The cells switch from aerobic respiration to fermentation. The lactic acid that accumulates as a waste product may cause muscle fatigue. But it is gradually carried away by the blood to the liver. Lactic acid is converted back to pyruvic acid by liver cells, but that process requires oxygen. Thus, fermentation in muscle cells results in an "oxygen debt" that is paid back when one continues to pant after having stopped exercising.

Exercising, like any other activity in life, is sometimes subject to short-cuts. These shortcuts in exercising are anabolic steroids; steroid use has become prevalent in all sports. Anabolic steroids are a group of powerful synthetic complex molecules that resemble the natural male sex hormones. All the steroids contain the following basic skeletal structure.

Hormones are chemical regulators in the body, which influence or control a wide range of processes such as growth, development and specialization of tissues, the reproductive cycle, and many aspects of behavior. Steroids (the steroid nucleus shown below. Note the methyl groups are a common feature of most steroids) are

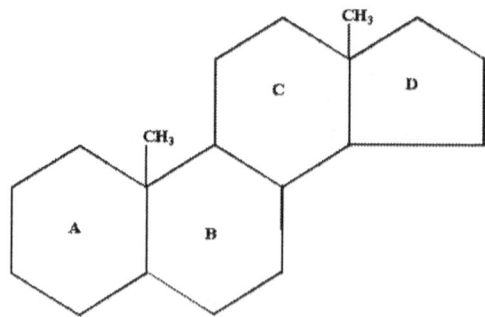

carried in the bloodstream and act as messengers. There are thousands of tiny steroid receptor sites in the cells, and the steroid molecules deliver many messages: among them the most important being to increase protein synthesis, and to increase creatine phosphate synthesis (creatine phosphate is the phosphage that the cells store as energy).

In short, it is easily understood that exercise and living in general are nothing more than a series of complex biochemical reactions.

The amount of calories one burns when exercising depends on the kind of activity one is engaged in. The more weight one moves - whether own body weight or a barbell- the more energy it takes. Here are some examples of the energy expenditure of various types of exercise.

ACTIVITY	CALORIES BURNED PER HOUR
Sitting	72-84
Walking (3 mph)	240-300
Calisthenics	300-360
Cycling (10 mph)	360-420
Jogging (5 mph)	420-480
Skiing	420-480
Running (5.5 mph)	600-660

Muscles, when exercised against adequate amounts of resistance, tend to become more efficient as well as stronger, seem to develop better tone and increased blood flow, and undoubtedly become less liable to suffer aches, pains, and injuries. In addition, exercising the muscles tends to counteract the process of muscular atrophy that inevitably happens as one ages. Exercising also stabilizes or lowers blood pressure over a period of time, strengthens the back thereby reducing the chances of lower back pain and other back problems, and increases the flow of blood to the skin, keeping it younger-looking and more flexible.

After the workout, she steps into the swimming pool to do few strokes. Since it is an indoor pool, the air is highly saturated with water vapor and imparts the strong odor of chlorine gas. The chlorine is usually added to water to keep it trouble free.

Chemistry in Service

Swimming Pool and Chlorine: Chlorine, under normal conditions, is a yellow green gas with bleaching action. Bacteria are the principle cause of unsanitary pool water. Very often people carry these microorganisms into the pool water creating unsanitary conditions. To disinfect or sterilize pool water, number of methods like electrolytic chlorine generation, ultraviolet sterilization, ozone oxidation, and reverse osmosis are being used on industrial scale, although these methods are gradually being adopted for residential pools.

Chlorine is by far the most popular disinfectant agent. It is available as gas, powder, and tablet and proven to be very effective and easy to use. Chlorine gas is generally restricted to public pools. Liquid chlorine in the form of sodium hypochlorite (NaOCl), dry chlorine in the form of calcium hypochlorite $(Ca(OCl)_2)$, and chlorinated isocyanurates are the major types used in residential pools.

When chlorine is added, some of it is used immediately to kill algae and bacteria, but some of it destroyed in this same process by algae and bacteria. The amount of chlorine used up in this manner is known as 'chlorine demand of water.' The remaining amount of chlorine left in the water is referred to as ' chlorine residual.' This free residual disinfectant keeps the pool sanitary.

Nitrogenous molecules, such as ammonia in the form of human wastes and fertilizers used near the pool also present in the pool water. Chlorine and ammonia combine to form chloramines (NH_2CL, $NHCL_2$, and $NHCL_3$), which cause burning eyes, skin irritation, and unpleasant odor - particularly pungent if the pH is low.

Chlorine residual should never drop below 1.0 ppm but it may go as high as 3.0 ppm. Chlorine residual is tested using orthotoluidine color test or Palin or DPD method or simple strip test.

Fifteen minutes in the pool, she is ready for hot and steamy shower to enhance further her muscle relaxation. Ten minutes in a shower does the job - she feels as if she is new. She walks through a giant drier to dry off and dresses up to go home. As she walks towards her car, she notices that the sky starts getting darker and feels like the thundershower is on its way. Not only that but the air smells quite distinct from normal dry air. Any way, she starts the car and tries to drive off the campus. Sure enough, the lightning and thunder from nowhere, without any slightest warning, start pounding.

Chemistry in Service

Thunder: Thunder is an electro-chemical phenomenon that usually followed by a heavy downpour. Inside a thundercloud, currents of air carry water droplets and hailstones upward, which rub against one another as they travel upward. A charge of static electricity is produced, and after awhile, a large amount of static electricity is present in the cloud. At the top of the cloud, there is a concentration of charge, while the bottom of the cloud is also charged because heavier particles and droplets tend to fall to the lower layers. It is then possible for a giant spark to be seen as electricity jumps between the top and bottom layers. The immense strike of thunder is due to the heated air expanding outward at great speed. Twigged lightning is produced when lightning jumps from the bottom of the cloud to the ground. A small fluffy cloud may carry anywhere from 100 to 1,000 tons of moisture: this is equivalent to the weight of 20 to 200 African elephants, and each year there are more than 16 million thunder storms around the world - there are about 2,000 in progress at any given time.

Mercury being a very dauntless person, she does not stop driving even in the roaring thunder and blitzing lightning. What you know! Within a short while, it begins to downpour. She could see that the water droplets as big as marbles start to bombard her car's windshield. She is no coward and has been in similar situation before many many times. She looks around, water, water everywhere. What is water?

Chemistry in Service

Water: Water (H_2O) (water is a common name, but its proper name is dihydrogen monoxide) molecule is formed by the chemical reaction between the hydrogen gas and oxygen gas in the presence of electric ark according to the following equation.

$$4 H_2 \text{ (gas)} + O_2 \text{ (gas)} \quad \xrightarrow{\text{electric ark}} \quad 2 H_2O \text{ (liquid)}, \quad \Delta H_f^0 = -285.8 \text{ kJ/mol}$$

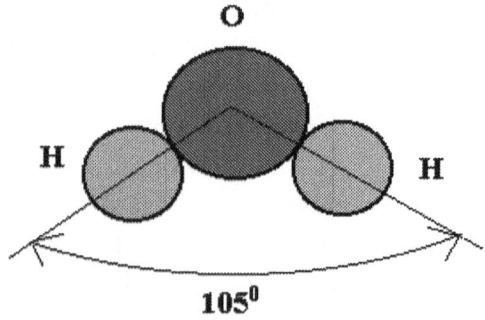

Two hydrogen atoms in water molecule are covalently bonded to oxygen. Water is the most unique liquid with very unusual chemical and physical properties. It is also the most abundant solvent on earth, and has immeasurable effects on our lives, from being an integral part of nutrition to the weather patterns. Approximately three-fourths of the surface of the earth is covered with water. Almost all the plants contain more than 50% of water. Nearly two-thirds, or 67%, of the weight of the human body comes from water. Many lower organisms consist of as much as 90% water. Air also contains water in the form of water vapor. The life on this planet could not survive without the water.

The chemistry of water is vast and interesting and here are some of its properties:

- Water expands upon freezing contrary to many other liquids.
- Density of solid water (ice), unlike most solids, is lower (0.917 g/cm^3) than liquid water (0.998 g/cm^3) due to formation of hydrogen bonds between the water molecules in ice. This is the reason why ice floats on water and lakes do not freeze in the wintertime - it is how nature protects marine lives in the lake in the wintertime.

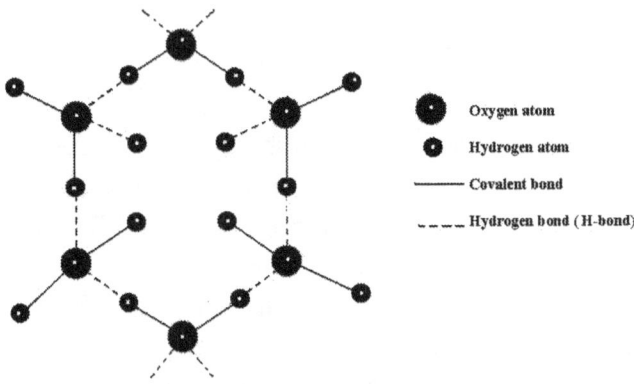

Oxygen atom

Hydrogen atom

Covalent bond

Hydrogen bond (H-bond)

- Water has the largest surface tension of any common liquid except liquid mercury.
- Water has an unusually large viscosity (viscosity is a measure of the resistance to flow).
- Water has an unusually high heat capacity (4.184 J deg^{-1} mol^{-1}); it takes more heat (4.184 J) to raise the temperature of 1 gram of water by 1^0 C than any other liquids.

Formation of Cloud: When a patch of warm ground heats up the air above it, a cloud begins to form. Since a bubble of warm air is created that is warmer than the surrounding air, it rises like a hot air balloon.

The bubble continues to rise, and as it does, it expands and cools. Eventually, the rising air bubble cools to below dew –point, the temperature below which water vapor condenses. In the cooled air, the water vapor then condenses on small particles of salt and dust thereby forming a cloud that hangs in the air. Air can be cooled below its dew -point by radiation, by mixing with cooler air, or by ascent in the atmosphere with resultant decompression. The rate at which water vapor is made available for the formation of cloud droplets is calculated from the Clausius-Clapeyron equation, which relates the saturation vapor pressure to Kelvin temperature (T):

$$P_s(T) / P_s(T_0) = \exp [- L_{vap} / R (1/T - 1/T_0)]$$

where $P_s(T)$ is the saturation or equilibrium vapor pressure at the temperature T, $P_s(T_0)$ is the saturation vapor pressure at reference temperature T_0, L_{vap} is the latent heat of vaporization and R is the gas constant (8.314 J/K mol).

The Clausius-Clapeyron equation enables us to calculate the increasing water or ice saturation ratio as appropriate for a bundle of cooling air containing a known quantity of water vapor. However, in order to determine the type of cloud (it's characteristics), it is necessary to consider the microphysical processes which control the formation and growth of the individual cloud droplets.

At this point, Mercury feels somewhat warmer inside the car because all windows in the car are closed tightly due to heavy rain. So, she orders the computer to roll down the windows a little bit. She feels that the air outside is already getting saturated with water vapor. As the humidity in the air starts rising up slowly, her windshield start getting foggier and foggier.

Chemistry in Service

Humidity: Humid air naturally contains more water vapor than average air. Thus, water vapor content of air is called humidity, and when it is expressed as a percentage of the maximum possible, it is called relative humidity.

Relative Humidity $= P_{H2O} / P^0_{H2O} \times 100$

where P_{H2O} is the partial vapor pressure of water and P_{H2O}^0 is the equilibrium vapor pressure of the water at the same temperature. Our daily living-comforts naturally depend upon the relative humidity as well as the temperature of our environment. When the relative humidity is above 80%, like in the hot summer days, perspiration does not evaporate quickly causing sticky and hot feeling. On the other hand, when the relative humidity is below 30%, like in the cold winter days, evaporation of the moisture from the body is rapid causing dry skin condition - nasal and mouth tissues dry out creating an easy passage for viruses to enter into the body. This may be why colds and flues are so common in the winter months.

Dew, fog, and smog (these are the colloidal solutions where the liquid water is dispersed in the gas medium) are able to form because warm air can contain more water vapor than cold air. Fog and mist are made up of tiny droplets of water that have condensed on extremely small particles such as salt particles in the air. Because fog droplets are so small, they remain in the air rather than falling to the ground. The sizes of droplets only measure about 1 to 10 microns (one micron is a millionth of a meter) across. According to international standards, the fog occurs when visibility is 600 ft. (180 m) or less. The longest occurring fogs, which can last for weeks, are found over the seas on the Grand Banks, Newfoundland in Canada.

Rain not only coming down heavily but also making the driving conditions most hazardous. She cannot see the road up ahead clearly due to poor visibility emerging from foggy windshield. In addition, fading away of headlights on her car as well as lights on streets becomes more eminent due to diffusion of lights that can be attributed to scattering effect. Mercury also starts feeling that her car is not gripping the road as it should be due to slippery conditions initiated by lowering of coefficient of friction. When the roads are wet, the water gets in between the tires and the road leading towards decrease in coefficient of friction or increase in slippery condition.

Chemistry in Service

Friction: Energy stored in all water surfaces, and particularly the energy that is stored in the surface of a water droplet (E_γ) of diameter d is given by,

$$E_\gamma = \pi\, d^2\, \tau$$

where τ is the coefficient of surface tension (the surface tension of a liquid can be defined as the amount energy necessary to stretch or to increase the surface by unit area), and is equal to about .075 J/m for an air-water interface at 0^0C and π (the ratio of the circumference of a circle to its diameter) is the constant the value of which is 3.1415629. The coefficient of friction is related to surface tension. It is defined as the constant characteristic of the surface in contact and is expressed through the following equation.

Magnitude of frictional force = coefficient of friction x magnitude of normal reaction

The coefficient of friction is an important phenomenon to know because when driving in a storm that involves either ice or rain, the coefficient of friction changes [when two solid bodies come in contact with each other at reasonably flat surfaces, a calculation can be made as to the frictional force resisting sliding motion of one surface across another]. For instance, the coefficient of friction is affected by varying degrees of surface smoothness and surface lubrication, and the speed of the sliding motion. The coefficient of friction for a smooth rubber tire on wet payment is about 0.5, as opposed to a smooth or grooved tire on dry payment, which is about 0.9, and a grooved tire on wet payment, which is about 0.8. Driving home in the rain is more dangerous because the coefficient of friction is always lower increasing the chances for sliding.

Scattering Phenomenon: Driving during poor weather conditions, in addition to slippery conditions, can also be attributed to low visibility caused by the scattering effect. Air, under normal weather conditions, is transparent to lights creating a good visibility. However, in the rain or fog, the concentration of water molecules in the air increases elevating the opacity of the air making it more susceptible to scattering of lights coming either from streetlights or oncoming cars. The scattering is actually done by the water molecules. This, in turn introduces what is known as 'poor visibility.'

The streetlights that are used today are an example of putting knowledge of chemistry at work. There is a minimum energy that is required to release an electron from the surface of an electrode. When a photon hits photoelectric surface its energy is absorbed by a surface electron. Part of the energy is used to release the electron, and the remainder transfers into kinetic energy of the ejected photoelectron. This phenomenon was described mathematically by Einstein as follows:

$$E_{photon} = W + 1/2 \, m \, v^2$$

where E_{photon} is the energy of the incident photon, W is the energy with which the electron is bound to the photoelectric surface, m is the mass of the electron, and v is the velocity of the ejecting electron. W, as defined above, is called work function of the metal, which is different for different metals and usually less than 10 eV. Because sodium (Na) has a low Work function of 2.29 eV, as compared to some other elements (Chromium 4.37 eV, Mercury 4.50 eV, Tungsten 4.58 eV, and Silver 4.73 eV) it is more efficient as a source of energy for our streetlights. Light is generated by electron bombardment of sodium vapor:

$$Na \quad + \quad energy \quad \rightarrow \quad Na^*$$
$$Na^{\cdot} \quad \rightarrow \quad Na + yellow\ light$$

where Na* is an atom in an excited electronic state. The resultant light that is given off is of great light intensity. In addition, yellow light has a longer wavelength than blue green light and due to that it is not readily scattered in fog. Many other substances also emit a visible light when they are excited with electricity. Both mercury and neon lamps also produce light when hit by electrons.

$$Hg \quad + \quad Energy\ (electrons) \quad \rightarrow \quad Hg^*$$
$$Hg^{\cdot} \quad \rightarrow Hg + blue\ green\ light$$

$$and \quad Ne \quad + \quad Energy\ (electrons) \quad \rightarrow \quad Ne^*$$
$$Ne^* \quad \rightarrow \quad Ne + orange\ red\ light$$

Mercury struggles through the heavy rain and storm to reach home. Getting home is not that bad considering the heavy storm, but to getting out of the car and getting inside the house is a great challenge as she quickly realizes that she had left her umbrella in the house. The storm is still at its peak and does not seem to budge an inch. She grabs her knapsack and tries to race to the door. But no luck, she gets soaked all over and even she hits the inconspicuous muddiest puddle on the street, making her cloths muddy and wet. She gets inside the house and turns the light switch on. Surprise! No electricity in the house due to heavy storm. So, she reaches for a flashlight, which she always keeps by the front door. Turns it on and feels so exhilarated to see the bright light from the flashlight spreading over the wide area like hawk, making the entire room visible to drenched and tired eyes.

Chemistry in Service

Battery and Electrical Energy: Electrical energy can be produced in several different ways. Some of the more common ways include chemical cells, electrical generators, heat, and the sun. Let us first look into how chemical cells produce electrical energy. Chemical cells are broken up into two categories wet and dry cells. Chemical cells contain three basic components. Each cell must have two electrodes (each electrode is different) that are placed in an electrolyte. One electrode collects positive charges from interaction with the electrolyte while the other collects negative

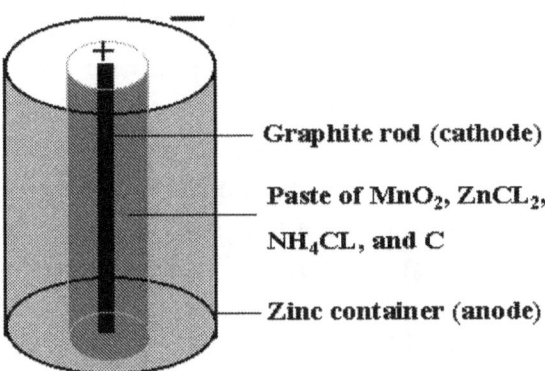

Graphite rod (cathode)

Paste of MnO_2, $ZnCL_2$, NH_4CL, and C

Zinc container (anode)

charges. These negative charges (electrons) move from the negative connection through an external circuit and reach the positive electrode. This movement of electrons produces an electrical current. There are several different types of dry cells. The ordinary dry cell (*Leclanche* cell) used in Mercury's flashlight shown above consists of zinc wall of the cell as anode and graphite rod in the center as cathode. The space between the electrodes is filled with moist paste of manganese dioxide (MnO_2), zinc chloride ($ZnCl_2$), and ammonium chloride (NH_4Cl). When cell operates, $Zn(s)$ at anode is oxidized liberating two moles of electrons according to

$$Zn(s) \rightarrow Zn+ + 2e^-$$

These electrons travel in the external circuit and enter at cathode where they reduce MnO_2 liberating ammonia (NH_3) according to

$$2\,MnO_2(s) + 2\,NH_4^+(aq) + 2e^- \rightarrow Mn_2O_3(s) + 2\,NH_3(aq) + H_2O(l)$$

The disadvantage of this type of battery is that graphite cathode gets insulated with ammonia gas reducing a voltage drop that results into shorter battery life. This problem can be avoided by using an 'alkaline' dry cell, in which NH_4Cl is replaced by potassium hydroxide (KOH). Although, alkaline battery is more expensive than *Leclanche* battery, it has longer shelf life and provides more current. On the other hand, the nickel cadmium cell uses nickel and cadmium as electrodes in electrolyses. This dry cell is unique because it has the ability to be recharged. The mercury cell is yet another dry cell. This cell uses mercuric oxide and graphite as one electrode and zinc and mercury as the other. This cell has the ability to hold full voltage almost throughout its lifetime.

Wet cells contain lead peroxide and lead electrode in acidic electrolyte. Wet cells are almost all rechargeable giving them a big advantage over most dry cells. Both chemical cells are used almost exclusively for batteries, both household and automobile.

Commercially available electricity is generated in the power plants in the form of AC (Alternate Current) electricity used in most households. This electricity is shipped to consumers via power lines. The household is charged for usage by a kilowatt-hour (kWh), which is 1000 watts of electricity. One kWh equals the energy used by a 100-watt light bulb for 10 hours, This energy in joules (J) is 3.6×10^6 J (= 100W x10 hx3600 (s/h) x (1 J/ 1W.s). One watt equals the energy consumption of one joule per second or watts = volts x amperes. The current is measured in amperes, a unit equal to the flow of 6.247 billion billion electrons per second.

To determine the electrical power of a device one must analyze three basic properties energy, current, and power. The energy measured in voltage. Then power equals energy multiplied by current (Power=energy x current).

Now, she is full of wet cloths and starts quivering. Hence, she slips into a robe and decides to wash off muddy clothes. What you know! Suddenly lights come on. Now she goes down to the basement and puts her wet cloths into the washing machine and adds some detergents to get rid off the mud and dirt, and starts the washing machine.

Chemistry in Service

Laundering Process: Laundering process is a cleaning technique to get rid off dirt, soil, stain or any other undesirable matter from the cloths. There are basically two types of cleaning; (a) wet cleaning that uses water and some kind of detergent, (b) dry cleaning that makes use of organic solvents instead of water and detergents.

In order to understand the chemistry involved in doing the whole laundry, one must look at the different steps involved in the laundering processes, which are:

1. Wet cleaning
   a. Soaps and detergents
   b. Chemical set-up
   c. Surfactant or detergent action
   d. Builders or phosphates
   e. Washer protection agents
   f. Bleaches
2. Dry Cleaning
3. Drying process - effect of surface area on the rate of drying.
4. Environmental effects

Most detergents are made of cleaning agents (anionic surfactants and enzymes), water softeners (sodium phosphates), washer protection agents (sodium silicates), fabric whiteners, colorants, and perfumes. The cleaning agents, just like soaps, are described as surface-active agents or surfactants and they permit water to penetrate the surface of an oil droplet and divide it into many fine particles, which become suspended in water (micelles). Each micelle traps an oil droplet and lifts it off thereby promoting water to penetrate into the minute pores of a surface and clean it throughout. Hand rubbing or agitation in a washing machine helps pull the dirt free. Since the soap molecules are active at these surfaces, they are called as **surfactants.** These molecules are also wetting agents. They tend to lower the surface tension of water and hence to improve water's ability to dislodge dirt is described as detergent **action.**

The main difference between soaps and detergents is that soaps create problems for laundering with hard water. When combined with salt-like end, they form a precipitate, which ends up as deposit on the clothes. To avoid this problem, cleaning agents in detergents are different than those in soaps. The hydrophilic end consists of a polar sulfonate group (SO_3^- Na^+)) and the hydrophobic end is made up of either long chain fatty acids or alkyl benzene:

$$CH_3 - (CH_2)_n - O - S(=O)_2 - O^- Na^+ \quad \text{Or} \quad R - C_6H_4 - S(=O)_2 - O^- Na^+$$

$$R = \text{straight chain alkyl group}$$

The above-mentioned detergents are anionic detergents because they have negative charged polar head group. In addition, there are also cationic, neutral, and amphoteric detergents, in which the polar head group is positive, neutral, or dipolar respectively.

$$CH_3 -(CH_2)_n - [- N^+ -(CH_3)_2 - CH_3] CL^- \qquad CH_3 - (CH_2)_n - C_6H_4 - O-(CH_2-CH-O)_{n'} -H$$

$$n= 15\text{-}17 \qquad\qquad\qquad n=7\text{-}11, n' = 5\text{-}10$$

cationic detergent   neutral detergent

$$CH_3 - (CH_2)_n - N^+ (CH_3)_2 - CH_2COO^-$$

$$n = 11\text{-}17$$

amphoteric detergent

There are also substances called **builders,** which chemically soften the water by binding to calcium and magnesium so that the detergent remains free to act as a surfactant. In addition, they also provide with other functions: (a) enhance the wetting effect and cleaning efficiency of detergents; (b) emulsify oily dirt by breaking it up and freeing it from the soiled surface; (c) suspend loosened dirt; (d) provide the alkalinity necessary for efficient cleaning without being hazardous; (e) buffer or maintain proper alkalinity in wash water; and (f) help reduce germ level on clothing. The major builders are sodium pyrophosphate and sodium tripolyphosphate, which are here shown binding with the metals in hard water.

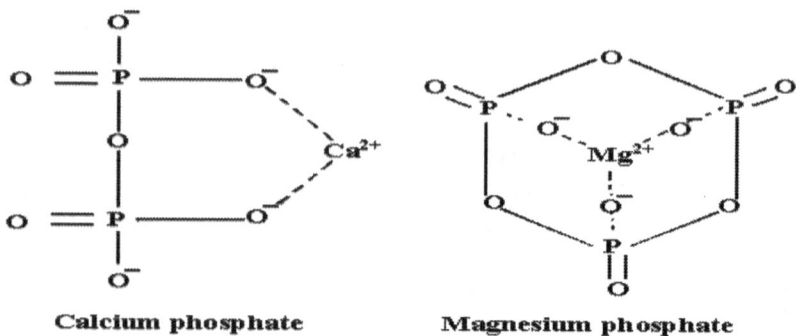

Calcium phosphate   **Magnesium phosphate**

The addition of these builders provides a better cleaning agent and at a lower cost.

Another substance that may be present in detergents is 'washer protection agents,' such as sodium silicates. These lubricate or protect the tank part of the washing machine.

Bleach is another substance used in laundering. Bleaches containing chlorine in the form of hypochlorites, such as, sodium hypochlorites (NaOCL) have been in use for many years. The chlorine acts as an oxidizing agent:

$$CLO^- (aq) + H_2O \quad \rightarrow \quad CL^- (aq) + 2OH^- (aq)$$

Due to strong odor of chlorine, more recently peroxides, especially, sodium perborate ($NaBO_3$) are being used. Hydrolysis of the perborate yields hydrogen peroxide (H_2O_2), which produces bleaching action.

$$NaBO_3 + H_2O \quad \rightarrow \quad H_2O_2 + \text{other products}$$

Fabric softeners give cloths a soft feel, which are cationic surfactants. They may be added to wash separately or may be incorporated in a detergent.

Enzymes are added to detergents to take care of specific types of stains; *proteases* are used to eliminate protein-based stains while *amylases* are used to get rid off starch-based stains.

Fabric brighteners are used to avoid yellow appearance of some white cloths. These are organic dyes that absorb ultraviolet light and fluoresce blue.

Cloth fibers swell in water. This may cause wrinkles, which are local distortions into the structure of the fiber, or even shrinking. Dry cleaning would prevent such a change since it uses nonpolar solvents that do not wet the cloth fibers. These solvents dissolve the nonpolar grease, freeing the soil particles to be removed by the detergents added. These are some of the halogenated hydrocarbons used in dry cleaning: carbon tetrachloride (CCL_4); trichloroethylene ($ClCH=CCL_2$); trichloroethane (H_3C-CCL_3); and chloroethylene ($CLCH=CHCL$)

When putting clothes in a dryer one must take into consideration of the effect of surface area on the rate of drying. - rate of drying increases as the surface area increases. Incidentally, the surface area can be increased by unfolding and uncreasing cloths before they are placed in the dryer. This eventually reduces the drying time propagating the reduction in drying cost.

Laundering process of course introduces few ill effects on our environment. For examples, the phosphates in detergents are insoluble in water (K_{sp} are very low) and have been linked to the overgrowth of algae in many lakes. 40% of all phosphates in these lakes come from detergents. This excess of nutrients (phosphorus) results in algae blooms or **eutrophication**. These algae end up using too much oxygen causing a **dead lake**.

Sewage treatment plants are thought of one of the ideas to reduce phosphates, but it is only in the experimental stage. Another idea is to replace a large amount of phosphates with sodium nitriloacetate (NTA), which works as a builder, but it is found that it binds to other toxic metal ions. Other substances have been tried, but they also suffer few side effects:

(a) Washing soda or sodium carbonate (Na_2CO_3) - has two side effects
   (i) Precipitate formation

$$M^{2+} + Na_2CO_3 \rightarrow MCO_3(s) + 2\,Na^+; \qquad M^{2+} = Ca^{2+}, Mg^{2+}$$

(ii) Produces high alkalinity-hazardous to small children

$$Na_2CO_3 + H_2O \rightarrow NaOH + NaHCO_3$$

(b) Silicates and soaps - form curd
(c) Metasilicates, citrates, perborates, and polycarbonates - produce high alkalinity and have poor biodegradability

To reduce the adverse effect on our environment, nowadays, all cleaning agents in the detergents are biodegradable as well as most are phosphate-free. All detergent bottles and carton boxes are all biodegradable. They are also made of 100% recycling plastic or paper, containing a minimum of 25% post-consumer recycled plastic or paper, some up to 50%.

After taking care of laundry, Mercury cooks some dinner as she is starving for the food due to all the strenuous exertion she went through. She takes the frozen TV dinner out of freezer, and puts into microwave oven for 15 minutes while she glances through her mail.

Chemistry in Service

Frozen Food: Frozen foods are one way of keeping the foods in their fresh condition -the foods are preserved for a longer period of time without the spoilage (undesirable products) either by reducing the rate of spoilage or inactivating the food enzymes. In general, the rate of reactions, chemical or biological, depends upon the temperature - it decreases as the temperature decreases. This principle is being exploited in frozen foods, which are usually kept in the freezer at low temperature enough to retard the rate of spoilage. In addition, free water becomes unavailable for growth and sustainment of microorganisms when water is converted to ice by freezing. Most microorganisms grow best in the range of 16- 38^0 C. Most pathogens cannot grow below 4^0 C. However, psychotropic bacteria thrive at lower temperatures and can grow slowly at temperatures down to 0^0 C and still below if water exits. Freezing alone is not enough to extend the shelf life of frozen foods. It must be combined with other methods, such as, concentration and dehydration, irradiation, pH control, chemical preservations, and effective packaging that include vacuum sealing using inert gases.

Microwave Oven: Microwave oven is an appliance that heats food by bombarding it with microwaves. These waves cause molecules in food to vibrate rapidly creating friction among themselves. This friction generates heat, which cooks the food. The microwaves are produced in a microwave oven by electronic vacuum tube called a magnetron, which is at a frequency of 2.45 GHz, with rated power output from 350 to 750 W. Microwave cooking generally takes much less time than the conventional ovens, like electric or gas because microwave ovens produce heat directly inside the food as opposed to outside as in the conventional ovens. Microwaves penetrate food to various depths depending upon the molecular makeup and thickness of the food. They pass through glass, paper, and almost all kinds of plastics and china. Metals, however, reflect the microwaves preventing them to enter the food, and hence metal containers should be avoided. In addition, if enough energy is reflected that may cause the damage to the magnetron. Microwave leakage from the oven may pose health-hazardous to people if they are close by, also reduces the efficiency of the oven. Hence, sealed doors are properly installed to prevent such leakage. Exposure to microwave power of 100 mW/cm^2 for several minutes is known to cause patho-physiologic effect in laboratory animals. These waves penetrate beneath the skin and may damage the tissue if the temperature rise is faster than the control mechanisms of the body.

By the time she finishes reading her mail, the dinner is ready. She sits down and eats the dinner. After eating dinner, she goes to the living room, turns the TV on to catch up with ' what is going on in the world today.' Today's two big news are: scientists have found a big ozone hole in the Antarctic region and thinks that it is going to have a devastating effect on our planet, and scientists are also deeply concerned with increasing concentration of carbon dioxide (CO_2) in the air, which is the main culprit in causing what is known as 'green-house effect.'

Chemistry in Service

Ozone and Our Environment: Ozone is a light blue toxic gas, used primarily in purifying drinking water, to bleach textiles, oils, and waxes, to deodorize air and sewage gases. Near the earth's surface it helps to create smog, which is pernicious to all living things. Contrary to that, it protects the living things when present in stratosphere.

Ozone (O_3), which consists of three oxygen atoms is the cousin to oxygen (O_2) or the allotropic form of oxygen, and is the 'savior' of our lives on this planet. The short-wavelength ultraviolet (UV) from the solar radiation has damaging effects on our lives, including, skin cancer, and mutations, etc. In addition, it also damages the plants and crops, parts of the food chain, and simple sea vegetation. It is anticipated that without ozone the life on Earth would gradually vanish. Then how ozone protects us?

Earth's atmosphere, three or four billion years ago, was mainly consisting of water and gases like methane and ammonia. At that time, there was not a trace of ozone. As the time passed by over the years, the concentration of ozone began to increase in the atmosphere mainly due to photosynthesis and photochemical decomposition of water liberating oxygen into the atmosphere. These oxygen molecules, when exposed to solar radiation with wavelength shorter than 260 nm in the upper atmosphere (stratosphere), broke up into two oxygen gas atoms, which in turn combined with oxygen molecule to produce ozone:

$$O_2 (g) + h\nu \quad \rightarrow \quad O(g) + O(g)$$
$$O(g) + O_2 (g) \quad \rightarrow \quad O_3(g)$$

where $h\nu$ is the energy of one photon. Ozone thus produced has the important photochemical property of absorbing solar radiation between the wavelengths of 200 nm and 300 nm to regenerate oxygen molecule and oxygen atom, which is reused in the regeneration of ozone as above:

$$O_3(g) + h\nu \quad \rightarrow \quad O_2(g) + O(g)$$

This is how ozone protects our lives by forbidding the UV radiation to reach the Earth's surface by absorbing itself.

In recent years, scientists have come to realize that ozone layer in the stratosphere is getting thinner and even there is a big hole created on Antarctic, which permit the passage of the harmful UV radiation to the Earth's surface. How could it happen? Who are responsible for this? The answers to these questions are none other than we, the human beings. We, as the citizens of the civilized and industrialized world, have created few chemicals that have a considerable love for ozone. These chemicals generally are known as a CFC's (chlorofluorocarbons) that are generally known as Freons and halons (fluorocarbons containing bromine). These are Freon 11 ($CFCl_3$), Freon 12 (CCL_2F_2), Freon 22 ($CHClF_2$), Freon 114 ($C_2Cl_2F_4$), and Freon 115 ($CClF_2CF_3$). They have found their ways in spray cans as aerosol propellants and in air conditioners as refrigerants due to their relatively inert, volatile, and easily liquefiable nature. In addition, they have also been used in Styrofoam. Solid particles and liquid particles can enter the respiratory tract and can cause serious toxicological problems. Once released into the atmosphere, they slowly but surely climb up into the stratosphere and decompose into free chlorine atoms with the absorption of UV radiation as follows.

$$CFCL_3 + h\nu \rightarrow CFCL_2 + CL$$
$$CCL_2F_2 + h\nu \rightarrow CCLF_2 + CL$$
$$CHCLF_2 + h\nu \rightarrow CHF_2 + CL$$
$$C_2CL_2F_4 + h\nu \rightarrow C_2CLF_4 + CL$$
$$\text{and} \quad CCLF_2CF_3 + h\nu \rightarrow CF_2CF_3 + CL$$

The chlorine atoms thus produced can do two things; (a) extract hydrogen from methane to produce hydrogen chloride (HCL) - an acid rain, and (b) engage in destroying the ozone molecules through the following mechanism:

$$Cl + O_3 \rightarrow ClO + O_2$$
$$ClO + O \rightarrow Cl + O_2$$

The net result is decreasing the concentration of ozone in the stratosphere:

$$O_3 + O \rightarrow 2O_2$$

The oxygen atom in the above reaction is supplied by the photochemical decomposition of ozone to oxygen molecule and oxygen atom. One chlorine atom can destroy thousands of ozone molecules since the radical (CL) is regenerated in the cycle.

A slightly different destruction mechanism suspected to prevail in creation of Antarctic ozone hole. It is assumed to take place at a particular time of year and in a place where no oxygen atoms present. In the first cycle, chlorine atoms react with ozone molecules producing chlorine monoxide and oxygen molecules. Chlorine atoms are then generated by reaction of chlorine monoxide with itself, which engages into the destruction of ozone molecule:

$$2\,CL + 2\,O_3 \rightarrow 2\,CLO + 2O_2$$
$$CLO + CLO \rightarrow CL_2O_2$$
$$CL_2O_2 + hv \rightarrow CL + CLO_2$$
$$CLO_2 \rightarrow CL + O_2$$
$$\overline{2\,O_3 + hv \rightarrow 3O_2\ (net)}$$

In addition to freons and halons, other families of radicals also have been suspected in destroying the ozone. For example, nitrogen oxide, which is produced by the supersonic aircraft flying in stratosphere:

$$NO + O_3 \rightarrow NO_2 + O_2$$
$$NO_2 + O \rightarrow NO + O_2$$
$$\overline{O_3 + O \rightarrow 2\,O_2\ (net)}$$

Recent studies have suggested that hydrogen oxide family may have involved in the ozone destruction cycle as indicated below:

$$OH + O_3 \rightarrow HO_2 + O_2$$
$$HO_2 + O_3 \rightarrow HO + 2O_2$$
$$\overline{2O_3 \rightarrow 3O_2\ (net)}$$

Now what is the remedy? Scientists and government in the United States are worrying together to solve this problem. The U.S. government already has banned the use of Freons in aerosol sprays and in air conditioners. Many industries have restrained themselves using in Styrofoams. In spite of these efforts, some other uses of Freons continues. How effective are these measures? Only time will tell !

Greenhouse Effect: This effect is the result of acting of Earth's atmosphere as the glass walls and roof of a greenhouse in trapping the heat from the sun. The atmosphere is mostly transparent to solar radiation, but has the tendency of strong absorption of infrared (longer-wavelength) from the Earth's surface. Much of this radiation bounces back to the earth surface and the net result is that Earth's surface receives more radiation than it would if the atmosphere were not cushioned between the sun and the Earth. The absorption of infrared radiation is dependent upon the small amounts of water vapor, carbon dioxide, ozone, methane, nitrous oxide, and other miner constituents of air, and presence of clouds. The greenhouse effect is most prominent at night and diurnal temperature usually ranges below 10^0C. Although the concept of greenhouse effect has been generally ascribed to the role of whole atmosphere in maintaining the Earth's surface warm, it has been increasingly associated with the contribution of carbon dioxide (a greenhouse gas), in the twentieth century anyway. It was pointed out long time ago in 1896 by Svante Arrhenius that increasing temperature at Earth's surface is due increase in carbon dioxide in the air produced by industrial combustion of fossil fuel. He also calculated that a doubling of concentration of carbon dioxide would raise the average temperature by about 5^0C. In addition to carbon dioxide, V. Ramanathan in 1975 suggested that greenhouse effect is further enhanced by the release of chlorofluorocarbons into the atmosphere, even though their combined concentration is less than 1 ppb by volume due to the spectral location of several of their absorption bands in 8-14 micrometer region, where the blackbody emission at terrestrial temperatures is somewhat high and the atmosphere is relatively transparent. Carbon monoxide, a much more abundant in air due to automobile exhaust and other combustion processes, does not play a direct role in enhancing the greenhouse effect because it does not have absorption bands in spectral regions where it can make a direction contribution. However, it does play an indirect role in amplifying the greenhouse effect by serving as a sink for hydroxyl radical, which acts as a catalyst in regulating the increase of nitrous oxide generated by use of chemical fertilizers and combustion processes, and methane (a natural gas) from biogenic and industrial production.

After hearing the news, she goes over notes she took in chemistry and biology lectures, but discovers that few things are missing. So she turns her notebook on that has a liquid crystal display, and connects to the NT server in the chemistry department, which is a part of wide area network in the school via wireless and retrieves the missing information. Since her notebook is equipped with multimedia technology, she has no difficulty in reading and listening to the lectures, which are usually recorded in multimedia format and stored on the department's server for reference purpose. While she is connected to the server, she also finishes her chemistry laboratory report and submits to professor via e-mail.

Chemistry in Service

Liquid Crystals: Today's notebook-sized PCs are equipped with flat-panel liquid-crystal display (LCD) screens instead of CRT (cathode ray tube), which are relatively heavy and bulky. Besides, LCD displays are digital rather than analog and due to that they are subject to much less distortion. In addition, the image is also sharp because LCDs do not use a relatively inaccurate electron gun. A special liquid is sandwiched between two plates. The bottom plate is reflective and the top plate is clear. The molecules in this liquid are normally aligned so that light passes through to the bottom surface and reflects back out through the top plate. Computer signals are used to rearrange the molecules in designated cells so that they block the passage of light. Today, Most LCD cells produce dark images on a silver background. But color LCD panels have been developed, it is anticipated that use of both monochrome and color LCD displays will grow rapidly in the future.

Most of the solids when melted pass directly to the liquid phase. However, some substances when melted do not pass to liquid phase directly, but first pass through an intermediate stage- *paracrystalline state* or *mesomorphic* - and then at a higher temperature undergo a transition to the liquid state. These intermediate states have been labeled as *liquid crystals,* since they display some of the properties of both liquid and crystalline states. Molecules with markedly unsymmetrical in shape tend to behave like liquid crystals. For examples, long-chain molecules in crystalline state may be oriented as shown in (a).

```
IIIIIIIIIIII    II IIIIIIIIIIII   I I I I I III I III \ / \ \ \ / \ /
IIIIIIIIIIII    II IIIIIIIIIIII   I I I I I I I I /\ / \ / / / \
IIIIIIIIIIII    I I III IIIIIIII  II I III II I I I ///\\\///\\
IIIIIIIIIIII    II IIIIIIIIIIII   II I   I I I II /\/\/\/\/\/\
 (a) crystalline   (b) smectic   (c) nematic   (d) isotropic fluid
```

When the temperature is raised, the kinetic energy may become sufficient just to disrupt only the bonding between ends of the molecules, but may not be sufficient to overcome the strong lateral attractions between the long chains. As a result of this, two types of anisotropic melts might be obtained - smectic and nematic as shown in (b) and (c). In smectic phase, molecules are still oriented in well-defined planes, and when stress is applied, the planes glide against each other. In nematic phase, the planar structure is lost, but the orientation is still preserved. On further increase in temperature, these anisotropic fluids become isotropic fluids (random orientation) as shown in (d).

By the time she finishes her homework, it is almost time for her to go to bed. However, she feels that she has a pounding headache, especially, caused by driving in the rain. Well, she must abate that pain before even she attempts to sleep. So, she reaches for an aspirin bottle in the kitchen cabinet, and swallows two tablets and about 8 x 10^{24} water molecules (250 ml), hoping that she will have a snugly sleep. What you know! She is already in deep sleep.

Chemistry in Service

Aspirin and The Stress: Frequently, the stress of the 'daily grind' results in the creation of the common headache. Dilated blood vessels in the brain and cranial region cause sensations to be felt in the receptors of the blood vessel walls. These sensations are perceived as 'nagging aches' and sometimes 'unbearable pain.' The protector of many afflicted people is often aspirin or some other type of analgesic such as acetaminophen, ibuprofen or codeine. The use of such chemicals to ease the body's condition demonstrates how chemistry is naturally associated with biology.

There are two types of pain, somatic pain (originating in the muscle or skeletal system) and visceral pain (arising in organs). Analgesics are used in treating somatic pain, such as a headache or muscle or joint inflammation, as well as fever. Headaches may be further classified as either intracranial or extracranial. Intracranial headaches may be serious, indicating a tumor or infection. The common headache falls within the extracranial category. Some forms of extracranial headaches treated with analgesics include: migraines, which are characterized by recurrent throbbing on the sides of the head and increased platelet aggregation; sinus headaches, distinguished by pain in the front of the head and around the eyes as the result of infection or allergy; psychogenic headache, which is often due to stress, anxiety, and depression and accounts for 90% of headaches. Furthermore, aspirins and analgesics inhibit prostaglandin synthesis. Prostaglandin's, a group of fatty acids containing 20 carbon atoms in a five-carbon ring and two tails, act on pain receptors by making them sensitive to incoming impulses.

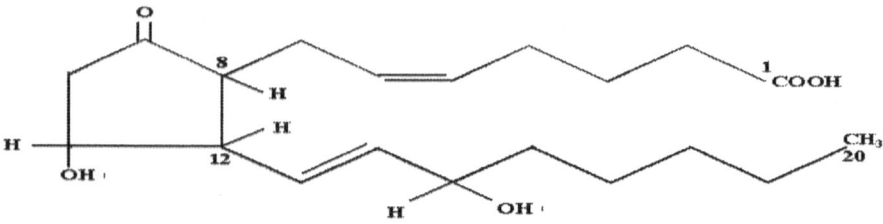

In the case of fever, aspirin reduces temperature by inhibiting prostaglandin production in the hypothalamus, the regulatory center within the body, which is responsible for the maintenance of homeostasis, as well as body temperature. As a result of analgesics, the hypothalamus implements the heat loss by increasing blood flow in blood vessels and induces sweating. Nonetheless, some analgesics are more effective for certain conditions than others. Only certain other medications may be ingested by some people as in the case of aspirin-sensitive patients. Analgesics, incidentally, are effective in treating minor aches and pains while severe pain is an indication of a serious condition and can often only be treated with narcotics, opiates, which attach to pain receptors and block pain impulses. Due to the severity of pain involved in migraine headaches, aspirin and other mild analgesics are ineffective and thus would need more potent drugs like opiates, for example codeine.

Basically, aspirin - acetylsalicylic acid - belongs to the group of salicylates, which are known for their analgesic effects. Aspirin is also referred to as ortho-acetoxybenzoic acid or salicylic acid acetate. The compound has a molecular weight of 180.16 (g/mol) and is only slightly soluble in water and soluble in alcohol and ether. Its structural formula is,

The inactive ingredients in aspirin tablet do not participate in the analgesic reactions. These include cellulose (a polysaccharide, corn starch), dyes such as red #40, yellow #10, or blue #1, magnesium stearate, hydroxypropyl methylcellulose, mineral oil, glycol, or wax such as *carnauba* wax. For the most part, these inactive ingredients act as binding agents and solidify the aspirin tablet, which in natural form is a white powder.

In addition, aspirin tablet, as purchased and consumed, is an ionic solid (which dissociates into ions upon ingestion) and is often combined with other substances along with inactive ingredients, such as calcium carbonate ($CaCO_3$), various magnesium compounds such as magnesium oxide (MgO) and magnesium carbonate ($MgCO_3$), and different sodium compounds like sodium bicarbonate ($NaHCO_3$) or sodium phosphate(Na_3PO_4). These substances are among the active ingredients which ameliorate the analgesic reaction, behave as neutralizing agents and are combined with the parent compound in aspirin to form buffered aspirin or effervescent-type aspirin. The buffered aspirin prevents gastric distress associated with dissolution into the stomach.

These buffering agents are designed to make aspirin dissolve more quickly in the stomach. The buffering agents neutralize the gastric juices (hydrochloric acid) and enzymes, such as pepsin, within the innately acidic stomach, thereby elevating the pH (pH is a measure of the acidity of a substance and is defined as the negative logarithm of hydrogen ion concentration, pH = -log[H$^+$]) of the stomach and making the aspirin easier to absorb with the walls of the stomach. An acid pH ranges from 1 to 7 and the base a pH ranges from 7 to 14. The pH of 7 is the neutral pH (see Appendix A for further details). Buffered aspirin, such as Bufferin, in addition to unbuffered aspirin also contains substances such as benzoic acid and citric acid. These two acids are preservatives and combine with the sodium in the form of sodium benzoate or salt are present in aspirin or more commonly in buffered aspirin or ant-acids, such as Alka-Seltzer. Unlike unbuffered aspirin, however, ant-acids contain sodium acetylsalicylate, which is not absorbed in the stomach, but instead is absorbed in the duodenum or small intestine.

Even more, aspirin tablets may also be coated in order to keep them from dissolving within the stomach and allowing them to be absorbed within the walls of the duodenum or the small intestine. Aspirins such as these are called 'enteric coated aspirin,' and are believed to reduce incidents of gastric bleeding, common to people taking unbuffered, uncoated aspirin. Enteric-coated aspirin is generally used in people using analgesics for a long duration. An example of this type of aspirin is Ecotrin. The ingredients including the parent compound are comparable to regular aspirin. Studies have proven that enteric-coated aspirin performs analgesically similar to regular aspirin, as it provides plasma salicylate levels not statistically different from plain aspirin. It should be noted, however, that enteric-coated tablets sometimes dissolve in the stomach or not at all. Additionally, aspirin may be timed-release and fabricated in order to prolong the pill's dissolution and reaction rate. Naturally, this form of aspirin is not effective in quick relief from the pain. More importantly, timed-release aspirin may cause gastric distress and bleeding. An example of timed-release aspirin is Bayer-timed release. It maintains similar ingredients to regular aspirin.

Other forms of aspirin include: calcium carbaspirin, containing calcium and urea (which is not ordinarily present in aspirin); chorine salicylate, liquid aspirin; magnesium salicylate; sodium salicylate; and salsalate. More often aspirin is combined with caffeine, also known as methytheobromine.

Most likely, caffeine is present in analgesics in the form of hydrobromide and sodium benzoate, which is a mixture of anhydrous caffeine, and sodium benzoate, as it is more soluble in water than regular caffeine. An example of an aspirin/caffeine combination is Excedrin and Anacin. Analgesically, caffeine affects mood and tightens distended blood vessels that may cause the pain in headaches. Also, aspirin may be combined with small doses of codeine, methylmorphine, and an opiate.

Codeine is present in analgesics as codeine phosphate structure($C_{18}H_{21}NO_3H_3PO_4.1/2H_2O$): Codeine is an alkaloid, which reacts as a base to form a salt. It has been implicated in the exacerbation of urticaria (hives).

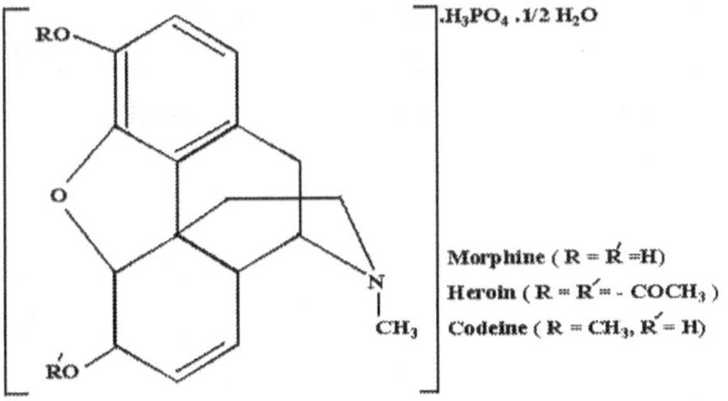

$.H_3PO_4 . 1/2 H_2O$

Morphine (R = R´ =H)
Heroin (R = R´ = - COCH₃)
Codeine (R = CH₃, R´ = H)

Commonly, codeine exists in combination with the popular analgesic like Tylenol containing acetaminophen, an aspirin-free (non-salicylate) compound ($C_8H_9NO_2$). Its structural formula is, HO- C_6 H_5 -NH-CO-CH_3 .

Acetaminophen (Tylenol) may often be found either alone or in combination with aspirin, as well. Regardless, the resulting tablet or caplet contains similar ingredients to plain aspirin. It has been noted that the standard minimum effective dose of both aspirin and acetaminophen is 325 mg. Like acetaminophen, ibuprofen (Advil and Motrin) is another effective analgesic used by aspirin-sensitive patients. Ibuprofen (+)-2-p-isbutylphenylproprionic acid has the structural formula:

$$CH(CH_3)_2 - CH_2 - \underset{\text{(benzene ring)}}{\bigcirc} - CH(CH_3)COOH$$

Similar to aspirin, ibuprofen often is combined with citric acid, sodium benzoate, starch and sucrose, glycol, hydroxypropyl methylcellulose, and stearic acid. It should be noted, however, that unlike acetaminophen, ibuprofen yields allergic reactions in people allergic to aspirin. Thus, people who are allergic to aspirin should take acetaminophen and not ibuprofen.

Aspirin, after it is consumed is rapidly absorbed within the walls of the stomach or duodenum. After absorption, aspirin is broken down into its basic component salicylic acid and acetic acid, as expressed by the chemical equation:

aspirin   + H_2O →   salicylic acid   + CH_3 COOH

                                                                  acetic acid

These components are then dispersed throughout the body. Salicylic acid and its metabolites bind to blood plasma proteins. Overall, aspirin is mainly metabolized by the hepatocytes in the liver. The essential metabolites are salicyluric acid, phenolic and acyl-glucronides of salicylate and gentisic acid. Nearly all of the aspirin is excreted by kidney. Acetaminophen and ibuprofen, similar to aspirin are rapidly absorbed from either the stomach or duodenum. The liver also mainly metabolizes them. Acetaminophen is metabolized into sulfate and glucuronide conjugates and excreted by the kidneys. Codeine, similar to aspirin is distributed to various organs and tissues within the body and is metabolized by the liver into morphine and norcodeine. However, it does not attach to blood proteins. It is excreted by the kidneys and appears in feces, as well.

Ultimately, aspirin and its counterparts acetaminophen and ibuprofen bear important 'pros and cons' to be considered. Principally, aspirin provides and irreversible affect on platelet aggregation, in other words, blood clotting, as well as the ability to synthesize prostaglandin, which are fundamental to the perception of pain - pain is the body's adaptive response that tells the brain that something is wrong within the body.

Ibuprofen has shown a reversible effect on platelet aggregation upon discontinuance of the drug. Meanwhile, acetaminophen does not influence platelet aggregation at all. Consequently, aspirin has been known to potentate excessive bleeding if blood vessel damage occurs. As a matter of fact, aspirin may incite latent ulcers and aggravate preexisting ones. Naturally, hemophiliacs and anemic are warned against consumption of aspirin. Common adverse side effects include stomach pain, nausea, vomiting, and heartburn. Frequently, buffered aspirin and ant-acid analgesic combinations increase blood pressure (hypertension) as a result of high sodium content. On the other hand, plain aspirin may ameliorate pregnancy-induced high blood pressure by reducing platelet production of thromboxane, which constricts blood vessels and raises blood pressure. Although aspirin consumption may prove beneficial in certain cases involving pregnant women, aspirin causes longer pregnancy, excessive blood loss during childbirth and increased infant birth weight. Even more, aspirin affects platelet aggregation in the fetus and new born and often prompts hemorrhaging. Thirdly, acetaminophen is recommended as an alternative to aspirin for pregnant women.

Other adverse effects of aspirin include increased uric acid retention of the blood. More importantly, aspirin hypersensitive people may display allergic reactions in the form of hives or asthma attacks due to prostaglandin inhibition by aspirin. Ibuprofen may yield similar reactions while acetaminophen does not. On the positive end, aspirin may reduce the risk of strokes and heart attacks due to its affect on blood clotting.

General References

Arms K. and Camp P.A., *Biology*, Saunders College Publishing Co., New York 1987.

Atkins P. W., *Molecules*, Scientific American Library, A Division of HPHLP, New York 1987.

Atkins P.W., *Atoms, Electrons, and Change*, Scientific American Library, A Division of HPHLP, New York 1987.

Barazini G.C., *Safe Practices in the Arts and Crafts*, A studio Guide.

Benigni A., *The New England Journal of Medicine*, 1989, 321, 357.

Bodner G.M. and Pardue H.L., *Chemistry An Experimental Science*, John Wiley and Sons, New York 1989.

Campbell N.A., *Biology*, The Benjamin/Cummings Publishing Company, Inc., Melno Park, CA 1987.

Complete Drug Reference, 1993 Edition, The United States Pharmacopeil Convention, 1992.

Considine D.W. and Considine G.D., *Encyclopedia of Chemistry*, Van Nostrand Reinhold Company, 1984.

Dennis A.S., *Weather Modification by Cloud Seeding*, Academic Press, New York, New York 1980.

Duckett K.W., *Modern Manuscripts*, American Association for State and Local History, U.S.A., 1975.

Getens R.J. and Stout G.L., *Painting Materials. A short Encyclopedia*, Dover Publications, Inc, USA, 1966.

Griffin *H.W., Complete Guide to Prescription and Nonprescription Drugs*, The Putnam Publishing Group, 1993.

Handbook of Nonprescription Drugs, American Pharmaceutical Association, Washington, D.C., 1986.

Helserman D.L., *Exploring Chemical Elements and Their Compounds*, TAB Books, PA, 1992.

Horgan J., *Gravity Quantized ?*, Scientific American, September, 1992.

Hart H., *Organic Chemistry - A Short Course*, Houghton Mifflin Company, Boston 1991.

Kauffman G.B. Ed. *The Central Science*, Texas Christian University Press, Fort Worth, Tx 1984.

Lafferty P., *Weather An Exploration of the Forces that Drive the World's Weather*, Crescent Books, ew Jersey 1992.

Lee J.C. and Bettelheim, F.A.,*Introduction to General, Organic and Biochemistry*, Saunders College Publishing, New York 1984.

Lehninger A.L., Short *Course in Biochemistry*, Worth Publishers, Inc., New York 1972.

Miller B.F., *The Complete Medical Guide*, Simon and Schuster, New York 1978.

Moore W.J., *Physical Chemistry*, Prentice-Hall. Inc., Englewood, N.J. 1962.

Metzger N., *Men and Molecules*, Crown Publishing Inc, N.Y. 1972.

Mayer R., *The Artist's Handbook of Materials and Techniques*, Ralph Mayers, USA, 1981.

Nourse A.E., *Universe, Earth and Atoms: The Story of Physics*, Harper and Row, New York 1969.

Nebel B.J. , *The Way the World Works*, Prentice-Hall, Inc., Englewood, N.J. 1981.

Physician's Desk Reference, Medical Economics Data, New Jersey 1993.

O' Dwyer J., *College Physics*, Brooks, Cole Publishing Co, California, 1981.

Orgel L.E., *The Origins of Life: Molecules and Natural Selection*, John Wiley and Sons, New York 1973.

Roberts M.T. and Etherington D., *Bookbindings and the Conservation of Books. A Dictionary of Descriptive Terminology*, Library of Congress, Washington, D,C, 1982.

River D.H., *Science*, Washington, D.C. 1992.

Stevens R.E., *University Archives*, Board of Trustees of the University of Illinois, Illinois, 1965.

Seidman A.H. and Flores I., *The Handbook of Computers and Computing*, Van Nostrand Reinhold Company, New York, New York 1984.

Seymour R.B. and Carraher, C.E., *Polymer Chemistry*, Marcel Dekker, Inc. New York 1988.

Sanders D.H., *Computers Today*, McGraw Hill Book Company, New York 1988.

Stryer L., *Biochemistry*, W.H. Freeman and Company, New York 1988.

Van Nostrand's Scientific Encyclopedia, D. Van Nostrand Company, Inc., Princeton, N.J. 1958.

Zimmerman D.R., *The Essential Guide to Nonprescription Drugs*, Harper and Row Publishers, New York 1983.

Appendix A
pH and Its Limitations

Mahadev Kumbar (1992)

The pH is a term used to describe the hydrogen activity in a solution. It is expressed as $pH = -\log a_H{}^+$, where $a_H{}^+$ is the activity of the hydrogen ion (1, 2). This definition suffers a considerable drawback because $a_H{}^+$ is not experimentally measurable(3). However, the pH is easily related to the standard free energy of the hydrogen ion with respect to a standard reference state (4,5). The hydrogen ion concentrations are generally expressed as 10^{-P}. The P in the exponent was called by Danish biochemist S.P.L. Sorensen as ' hydrogen ion exponent' and replaced it with pH. Therefore, the hydrogen ion exponent (pH) of a solution means the Briggs logarithm of the reciprocal value of the hydrogen ion concentration (1,2,6).

The purpose of defining the pH as the negative logarithm of hydrogen($[H^+]$) ion concentration or hydronium ion($[H_3O^+]$) concentration ($pH = -\log_{10}([H^+]) = -\log_{10}([H_3O^+]) = \log_{10}(1/[H_3O^+])$) is to avoid working with negative exponents appearing in the hydrogen ion concentration. In addition, this transformation leads toward easily manageable positive numbers. Such procedure is essential since hydrogen ion concentrations are usually very small in their magnitudes. When Sorensen in 1909 proposed the transformation of hydrogen ion concentration in terms of pH, it was generally thought that strong electrolytes, like HCl, partially ionize in dilute solutions. Besides, most likely Sorensen did not have any idea about the limits of hydrogen ion concentrations that one-day might exceed the limits set by him. This procedure works very well as long as $[H^+] < 1$ M and fails if $[H^+] > 1$ M defeating the basic principle underlying the pH definition. The currently accepted pH range ($0 =< pH =< 14$) or the hydrogen ion concentration range ($1 \times 10^0 >= [H^+] >= 1 \times 10^{-14}$) seems insufficient to include the pH of those solutions that have $[H^+] > 1$ M. There are many examples in chemistry where $[H^+]$ exceeds 1 M; some of the most common acids and bases kept in the laboratory in dilute forms have 3 M or 6 M concentrations and have greater than 3 M or 6 M concentrations in concentrated forms (Table I).

Table I. Concentration of some common laboratory acid and base solutions.

Reagent	Molarity	pH*	TpH
Hydrochloric acid (conc.)	12	-1.08	1.92
Hydrochloric acid (dil.)	6	-0.78	2.22
Nitric acid (conc.)	16	-1.20	1.80
Nitric acid (dil.)	6	-0.78	2.22
Sulfuric acid (conc.)	18	-1.26	1.74
Sulfuric acid (dil.)	3	-0.48	2.52
Sodium hydroxide (conc.)	14	15.15	18.15
Sodium hydroxide (dil.)	6	14.78	17.78

* calculated based on the assumption that the solutions are completely ionized including sulfuric acid.

Assuming that these solutions still retain their complete ionization properties at these levels of concentrations, the pH values of strong acids become negative numbers defeating Sorensen's intention. On the other end of the scale, the pH values of strong bases, like NaOH (14 M), take on positive numbers in conformity with the pH definition but exceed currently set upper limit of 14 (see Table I). **Hence, the first limit on the pH is embedded in the outer limits imposed on the pH scale.** Therefore, it appears that a more general approach to cover a wider range of pH or hydrogen ion concentration than currently accepted practice may be necessary.

Reformatting the pH also seems to be favored by the human nature; it is my experience teaching chemistry for many years that most students do not quite tend to grasp or remember either the acid range ($0 =< pH < 7$) or the base range ($7 < pH =< 14$), or even to that matter, the value of the neutral pH = 7. One of the difficulties, as I see it, might be due to the oddity of these numerical values. However, if these values are redefined in an 'easily rememberable' forms, then I think that students will have no difficulty in remembering these ranges. The neutral solution is often thought of as a 'prefect' or a 'balanced' solution in the sense that hydrogen ion concentration is perfectly balanced by the hydroxide ion concentration ($[H^+]$ = $[OH^-]$ = 1×10^{-7} M). If this solution is assigned the value of 10 instead of 7 (I am not proposing here to change the concentrations of $[H^+]$ and $[OH^-]$ in pure water at 25^0 C from 1×10^{-7} M) , then I am confident that students will remember it without any difficulty because the number 10 is used to display the prefect tendency in many instances in our daily lives. For examples, the scale between 0 and 10 is used in rating various sports activities including many Olympic games, many computer magazines use the scale of 0 to 10 to rate softwares, and many radio stations use the scale between 0 and 10 to forecast the sunburn index in the summer time, the SI system is based on 10, etc. In view of these examples, it may be worthwhile to redefine acid and base ranges respectively as between $0 =< acid < 10$ and $10 < base =< 20$ to synchronize with common practice and also to give students a better way of remembering the pH scale without any struggle.

Therefore, on the bases of the above argument, we may now attempt to transform the pH and pOH using the following formulas, and call the transformed pH as TpH (T for Transformed) and that of pOH as TpOH:

$$TpH = pH + 3.0 = -\log_{10} \{ [H^+] \times 10^3 \}$$

and $TpOH = pOH + 3.0 = -\log_{10} \{ [OH^-] \times 10^3 \}$. (1)

Of course, the concept involved in this transformation is not new but similar to that of Kelvin temperature scale. Thus, transforming the pH in the above form covers $[H^+]$ up to 1×10^3 M. If $[H^+]$ exceeds this value, the TpH like pH, also becomes negative. Let us hope that we never approach that limit. On the base side, the $[H^+]$ may be extended up to 1×10^{-17} M to make the scale even on both sides of the neutral TpH = 10.0. The redefined acid range (0 =< TpH < 10) and the base range (10 < TpH =< 20), like neutral pH (=10), are also in an 'easily rememberable' forms due to metric nature of the numbers 10 and 20. Hence, redefining the pH as above makes the pH scale as a subset of the TpH scale. The extended portions of the TpH scale may be distinguished by labeling them differently from the labels in the pH scale; The range 0 =< TpH =< 2 (-3 =< pH =< -1) may be labeled as 'superacid' range while the range 18 =< TpH =< 20 (15 =< pH =< 17)) as 'superbase' range. The sum of TpH and TpOH, like the sum of pH and pOH (pH +pOH =14), also becomes a constant number, that is,

TpH + TpOH = 20. (2)

The detail comparison of pH and TpH scales is summarized in Table II.

Table II. Comparison of pH and TpH scales.

$[H^+]$	$[OH^-]$	pH	pOH	TpH	TpOH	Solution pH Scale	TpH Scale
10^3	10^{-17}	-3	17	0	20		
10^2	10^{-16}	-2	16	1	19		superacid range
10^1	10^{-15}	-1	15	2	18		
10^0	10^{-14}	0	14	3	17		
10^{-1}	10^{-13}	1	13	4	16		
10^{-2}	10^{-12}	2	12	5	15	acid range	acid range
.	.	.	.	.	.	.	.
.	.	.	.	.	.	.	.
10^{-7}	10^{-7}	7	7	10	10	neutral	neutral
.	.	.	.	.	.	.	.
.	.	.	.	.	.	.	.
10^{-12}	10^{-2}	12	2	15	5	base range	base range
10^{-13}	10^{-1}	13	1	16	4		
10^{-14}	10^0	14	0	17	3		
10^{-15}	10^1	15	-1	18	2		
10^{-16}	10^2	16	-2	19	1		superbase range
10^{-17}	10^3	17	-3	20	0		

The second limitation is manifested in the way many general chemistry text books write the formula : $pH = -\log_{10}[H^+]$. The hydrogen ion concentration is usually expressed in mol/L. It is not clear what happens to the unit mol/L when logarithmic transformation is applied to the $[H^+]$? As we know that the logs are operative only on pure numbers and due that the unit mol/L is simply neglected.

In addition, the hydrogen ion concentration is calculated by the formula, $[H^+] = 10^{-pH}$. In order to take the power of 10 the number (pH in this case) must be a pure number, as is the case here. However, the answer derived from this formula, namely $[H^+]$, is expressed in mol/L. Where the unit mol/L comes from? It is not part of the equation $[H^+] = 10^{-pH}$. In reality the unit, mol/L, is there but is hidden or not obvious.

Therefore, to avoid such confusion, the formulas for pH and pOH may be written as,

$$pH = -\log_{10}\{ [H^+] / (mol/L) \}$$

and $$pOH = -\log_{10}\{ [OH^-] / (mol/L) \}. \qquad (3)$$

Thus, dividing the $[H^+]$ or $[OH^-]$ by mol/L produces a pure number because hydrogen ion or hydroxide ion concentrations are expressed in mol/L. The above formulas also correctly produce expressions for $[H^+]$ and $[OH^-]$:

$$[H^+] = 10^{-pH} \; mol/L$$

and $$[OH^-] = 10^{-pOH} \; mol/L. \qquad (4)$$

Example. Let $[H^+] = 2.5 \times 10^{-5}$ mol/L. Then pH $= -\log \{ 2.5 \times 10^{-5} (mol/L) / (mol/L) \} = -\log (2.5 \times 10^{-5}) = 4.6021$. This pH can be used to reproduce the original hydrogen ion concentration: $[H^+] = 10^{-pH}$ mol/L $= 10^{-4.6021}$ mol/L $= 2.5 \times 10^{-5}$ mol/L .

The transformed formulas (Equations 1 & 2) set the upper limit for pH as 10^3 mol/L and lower limit as 10^{-17} mol/L. At this point, let us ask ourselves a question: is it possible to detect the hydrogen ion concentration as low as $\sim 10^{-17}$ mol/L or as high as 10^3 mol/L? Only experimentalists can answer this question.

References

1. Sorensen S.P.L., *Compt. Rend. Lab.* Carlsberg, 1909, 8,1.
2. Sorensen S.P.L and Lindstrom Lang K, *Compt. Rend. Lab.* Carlberg, 1924, 15, 40, 1924.
3. Clerk WestCott C., *Ph Measurements*, Academic Press, N.Y. 1978.
4. Moore W.J., *Physical Chemistry*, Prentice-Hall International, Inc, Englewood Cliffs, N.Y. 1962.

5. Sorensen S.P.L., *In Source Book in Chemistry* 1900-1950, Edited by H.M. Leicester, Harvard University Press, Cambridge, Mass. 1968, p 16-19.

6. Harned H.S. and Owen B.B, *The Physical Chemistry of Elelctrolytic Solutions*, Reinhold Publishing Corp., N.Y. 1958.

Appendix B
Fullerene I. Computer Simulation of C_{60} Molecule

Sujata Kumbar and Mahadev Kumbar(1994)

Buckminsterfullerene (in honor of Buckminister Fuller who invented geodesics (1,2)) or simply known as buckyball consisting of sixty carbon atoms intertwined into two types of rings - pentagons and hexagons, molded into the shape of a soccer ball was the dream- molecule of Chapman and his associates (3,4), was brought down from heaven to earth by Kratschmer *et al* (5), and finally gave the life and the physical appearance and even the most exotic name - buckminsterfullerene - by Smalley (6) and Kroto (7), who demonstrated the craftsmanship of chemists and the elegance of the inner beauty of the molecular world. Of course, such striking shapes are not new to mankind and have been known to mathematicians (8,9,10) for centuries. Since the dawn of buckyball-civilization, the news media has been filled with its glory (11), and the literature has been exploded with vast number of articles to characterize its physical appearance and to assess its behavior in the real world by using all possibly known tools to chemists. Fullerenes (a class of molecules akin to buckyball) have already been discovered in nature - in ancient Russian rock (12). The methodologies employed in characterizing the fullerenes can be classified into two broad categories; (a) methods devoted to structural determination, and (b) methods directed to describe their physical and chemical properties. The first type of methods include X-ray diffraction studies on metal ion intercalated inside and outside the cage of the molecule and also complexed with non-polar molecules (13-21), and solid state NMR studies (22,23). The second type of methods embrace optical spectra and vibrational spectra(IR and Raman) (24-30). A considerable effort has also been extended to understand the pathway leading towards the complete formation of fullerenes (31,32). Number of other akin molecules have also been synthesized and characterized (17,22,26,27,33-36). The superconducting properties of some fullerides (fullerene anions) have also been investigated (37,38). Such classes of molecules are not alone but have a family of precursors known as corannulenes (39-42). In addition to these experimental studies, some theoretical calculations also have been carried out (43-47, see also the reference at the end of Table I).

Table I. Bond length (A^0) in C_{60} molecule from various sources.

Compound	$C_5 - C_6$	$C_6 - C_6$	Method	Reference
C_{60} -Cyclohexane	1.51	1.44	X-ray	14
C_{60} - Osmylated	1.432	1.388	X-ray	16
C_{60} - Pt derivatives	1.45	1.39	X-ray	19
C_{60}	1.45	1.40	NMR	24
C_{60}	1.42	1.42	EXAFS*	21
C_{60}	1.36-1.45	1.34-1.38	THEORETICAL	43,44[#]
C_{60}	1.42	1.394	Huckel	This work

* Extended X-ray Absorption Fine Structure.
# See also Tables 6.2 and 6.3 (White,C.T.; Mintmire,J.W.;Mowrey,R.C.; Brenner,D.W.; Robertson,D.M.; Harrison, J.T.; Dunlap, B.I..in *Buckminsterfullerenes*, Billups, W.E.;Ciufolini,M.A.; Eds., VCH Publishers, New York 1993, chapter 6) which summarize the results of numerous theoretical methods.

The role of carbon in this nature is simply amazing. It has given us all sorts of structures in various sizes and shapes embedded with beauty and tantalizing properties. Its evolution, after the bigbang, must be pretty intriguing. Fascinating account of the synthesis of carbon in the galaxy has been put into prospective by Hare and Kroto (48).

The aforementioned investigations were able to bring out some characteristics and dynamics of the buckyball. All these studies agree that buckyball is a rigid hollow spheroidal molecule with 12 pentagons and 20 hexagons - pentagon is always surrounded by 5 hexagons and hexagon is surrounded by alternating 3 pentagons and 3 hexagons in the shape of truncated icosahedron; a polyhedron known in this name to mathematical world- and do not quite seem to agree on the bond lengths and the radius. As many of the properties critically depend on these two parameters, we have undertaken this investigation to examine further the dependency of number of properties on these two parameters. To do so, we have built a molecule with the aid of the computer (the computer program was written in BASIC language by us) and minimizing technique. Our calculated properties definitely tend to compliment the existing properties derived from various other methods, and further appear to shed a light on the behavior of some of the properties.

As discussed above, great attention has been focused on the C-C bond length in C_{60} molecule due its primary role in dictating shape, size, and eventual evolution of physico-chemical properties. Variety of experimental as well as some theoretical studies so far conducted differ in assigning this length, although, there is some general agreement about the order of magnitude (see Table I). As a further exploration, we have calculated this bond length using Huckel molecular orbital method. Considering the simplistic nature of this method, it remarkably has yielded bond lengths that are in the range predicted by other methods.

Computer Simulation Procedure

There are numerous ways, as one can imagine, building C_{60} molecule on the computer. Basically, our procedure was first to build a half-molecule consisting of 6 pentagons and 10 hexagons, and then join two-halves (dimerize) together to create an entire molecule. Initially, the local coordinates of the pentagon and the hexagon were determined by using the coordinate system shown in Figure 1(a) and 1(b).

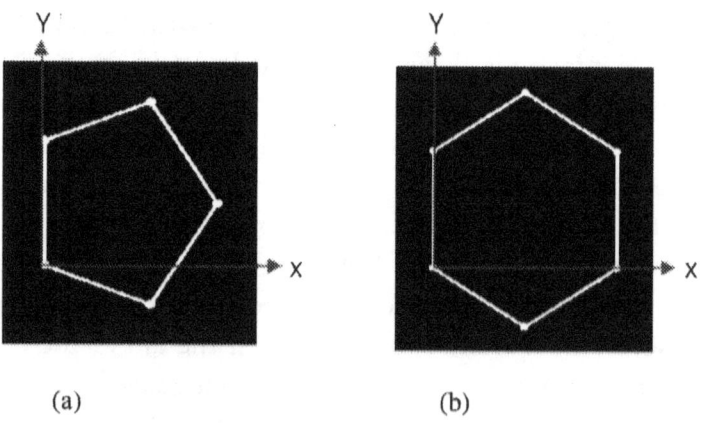

(a)   (b)

Figure 1. Local coordinate system for pentagon (a) and hexagon(b). The origin (0,0,0) is located at one of the carbon atoms. The y-axis is placed along C-C bond, the x- and z-axis are placed in such a way as to complete the right-handed coordinate system. The bond angles are 108^0 and 120^0 respectively in pentagon and hexagon.

The standard bond angles and bond lengths (49) were used in determining the local coordinates and eventually building the molecule. Subsequently, the C_{60} molecule was also built by varying the bond length between the standard single bond length and the standard double bond length. We started building the molecule around the pentagon. We attached the global coordinate system to pentagon in such a way that C_1-

atom was placed at the origin of the coordinate system, Y-axis along the C_1- C_5 bond, and X- and Z-axis were placed in such a way as to complete the right-handed coordinate system (Figure 2 (I)).

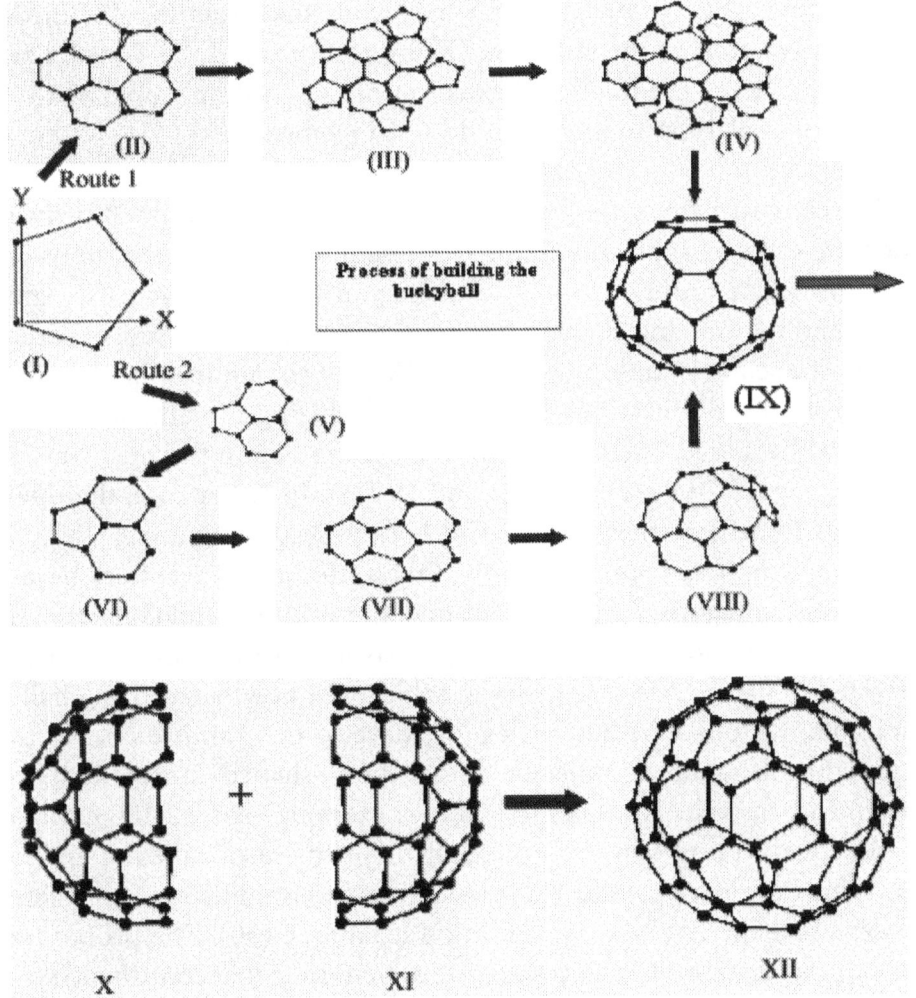

Figure 2. Global coordinate system (I). The C_1 -atom of the pentagon is placed at the origin. Y-axis along C_1 - C_5 bond, X-axis in the plane of

the pentagon, and Z-axis in such a way as to complete the right-handed coordinate system. Structures(II), (III), and (IV) indicate the first route and structures (V), (VI), and (VII) the second route of building the bowl-shaped molecule(IX). The structure (X) represents the rotation of the structure (IX) by 90^0 around Y-axis while the structure (XI) indicates the rotation of the structure (X) by 180^0 around the Y-axis. The structure (XII) is the final conformation of C_{60} molecule obtained by fusing both halves(structure(X) and (XI)) together.

The construction of the half-molecule may be accomplished by two alternate routes; (a) placing all the required hexagons and pentagons in the same plane around the central pentagon (Figure 2 (II), (III), and (IV)) and minimizing all the necessary dihedral angles leading towards a bowl-shaped half-molecule (Figure 2(IX)); or (b) adding one ring at a time around the central pentagon and minimizing the dihedral angle between the added one and the preceding one before the next ring is added (Figure 2 (V), (VI), (VII) and (VIII)). In either case, the final goal was to achieve the bowl-shaped half-molecule (Figure 2 (IX)), a molecule akin to *corannulene.* In the first route, all the necessary hexagons and pentagons were place around the central pentagon leading towards a flat molecule (all the rings existing in the same plane) as depicted in Figure 2(IV). In order to achieve the bowl-shaped C_{30} half-molecule, various dihedral angles $\{\varphi_i\}$ have to be minimized so as to mold this flat-shaped molecule into a bowl-shaped molecule. This means that the valence angles $\{\theta_i\}$ (see for example Figure 2(II)) have to be set equal to zero values through the adjustment of dihedral angles $\{\varphi_i\}$. Thus, we have carried out the simultaneous adjustment of these angles using the Powell's minimization technique (50). The procedure for transformation of local coordinate system to global coordinate system is described elsewhere (51,52).

In the second route, two hexagons were first placed around the central pentagon (Figure 2 (V)) and the torsional angles φ_1 and φ_2 were minimized to give the structure VI. A further addition of hexagons and minimization lead to the structure VII completing the first round of addition of hexagons. In the next round, alternating pentagons and hexagons were placed (Figure 2 (VIII)) and corresponding dihedral angles were minimized leading towards the bowl-shaped structure IX. The main difference between the first route and the second route is that all fifteen dihedral angles are needed to be minimized simultaneously in the first route, while only two dihedral angles at a time are needed to be minimized in the second route.

Construction of the C_{60} molecule from two C_{30} - halves was (dimerization process) proceeded as follows. The half-molecule shown in Figure 2 (IX) was first rotated around the Y-axis by 90^0 to give the molecule shown in Figure 2(X). The latter molecule was rotated by 180^0 along the Y-axis to give its mirror image (Figure 2(XI)). Finally, both halves were joined together, again, using the Powell's minimization procedure (50) to yield the final spheroidal C_{60} molecule (Figure 2 (XII)).

Results and Discussion

(a) **Dihedral angles.** When we optimized all the 15 dihedral angles, we found that they fell into two distinct categories; those between C_5- C_6 had the same value of 37.36^0 and those between C_6-C_6 had the same value of 41.76^0 (the convention adopted here is that these angles are zero when both rings lie in the same plane). Thus, two distinct values that are obtained might be attributed to the way the pentagons and hexagons are joined together at their respective junctions propagating the difference in bond lengths or vice versa in agreement with numerous other studies (see Table I).

(b) **Radius and Volume.** Variety of studies (Table I) have indicated that C_5-C_6 and C_6 - C_6 bonds do not have the same value but range from 1.34 A^0 to 1.51 A^0. The value of 1.34 A^0 is very close to the standard double bond length of 1.33 A^0, and 1.51 A^0 is close to the standard single bond length of 1.54 A^0. Therefore, we have investigated the entire range of values from double bond to single bond to assess the effect of this variation on the radius and the volume of C_{60} molecule.

Table II. Radius (A^0) and Volume of C_{60} molecule.

C - C bond length (A^0)	Radius(A^0)	Volume(A^{03})	Reference
1.33 (std. double bond, sp)	3.29	149.89	this work
1.35	3.34	156.76	this work
1.397 (std. aromatic, sp^2)	3.46	173.71	this work
1.40	3.47	174.83	this work
1.45	3.59	194.24	this work
1.50	3.72	215.03	this work
1.54 (std. single bond, sp^3)	3.82	232.70	this work
	3.53		18
	3.55		49
	3.55		22

It is seen (Table II, Figures 3 and 4) that both radius and volume (calculated by using the formula $4/3\ \pi\ r^3$) follow linear trend. The radius increases linearly as the C-C bond length increases, which is fitted to the following equation:

Radius (A^0) = - 0.06 + [2.52 x C-C bond length]   ($R^2 = 1.0$)   (1)

On the other hand, the volume also increases as C-C bond length increases but in somewhat an exponential fashion (Figure 4).

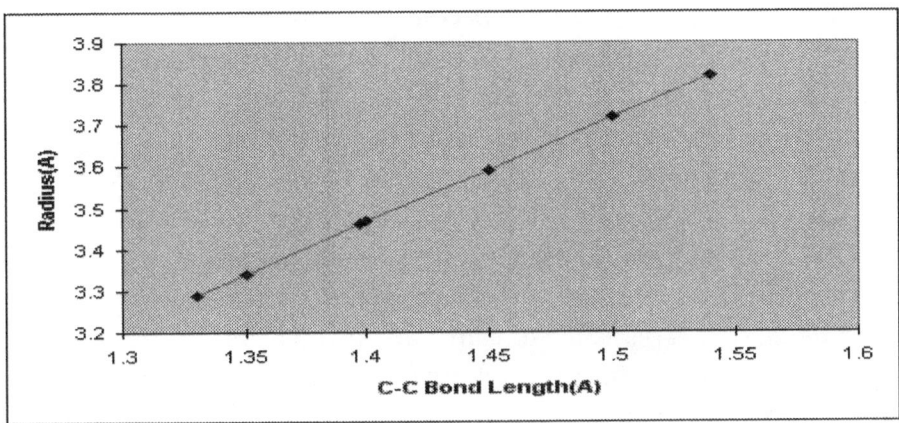

Figure 3. Plot of radius(A^0) of C_{60} molecule as a function of C-C bond length(A^0) for various values shown in Table I.

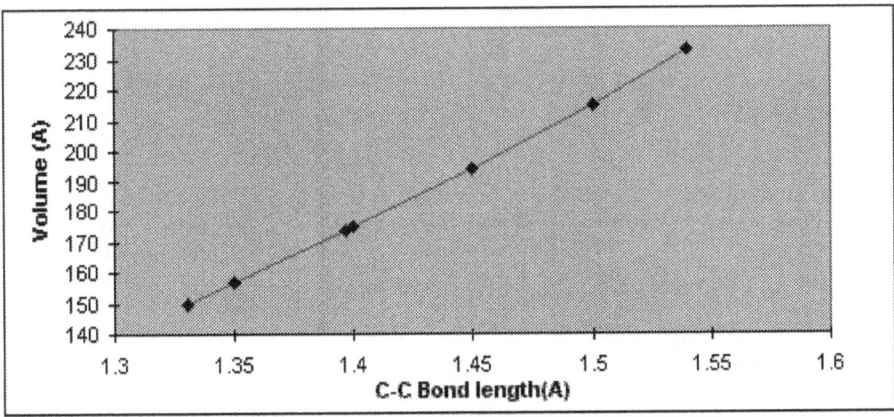

Figure 4. Plot of volume (A^{03}) of C_{60} molecule vs. C-C bond length (A^0) for various values listed in Table II.

The choice of increment in bond length (Table II) is of course arbitrary. In calculating the radius and the volume, we have set both C_5 - C_6 and C_6 - C_6 lengths equal to one other - otherwise generating the C_{60} molecule with two different bond lengths becomes a difficult, if not impossible, task. Nevertheless, these values for the radius fall within the range obtained by numerous experiments (Table II). Since there is a close match between our calculated values and experimental values, the justification for setting both bond lengths to the same value appears to be valid.

Smalley (17) has synthesized inner cage molecule with lanthanide ion in it. In view of this, it seems appropriate to ask the question, such as, *what is the maximum size of an ion or a molecule that can be fitted inside the cage?* It is plausible to gain some insight into this problem by knowing the radius of the C_{60} molecule and van der Waals radius of the carbon atom. For example, if we take the radius as 3.46 A^0 for standard aromatic bond length (Table II) and 0.77 A^0 for van der Waals radius for carbon atom, we may estimate the radius of inscribed sphere as about 3.08 A^0 (volume = 12.9 A^{03}). In Table III, we have listed the volumes of numerous simple molecules of different sizes and shapes.

Table III. Dimension of various small molecules. d_a and d_b are the diameters of the major and minor axes. Volumes are calculated assuming the shapes listed in the volume column for each type of molecule. Mathematical formulas used in these calculations may be found in reference 56.

Type	$d_a (A^0)$	$d_b (A^0)$	Volume(A^{03})
	(major)	(minor)	
Linear molecules			*cylinder*
C_2H_2	3.94	1.54	7.34
Cl_2	3.96	1.98	12.19
CO	2.63	1.54	4.89
CO_2	3.78	1.46	6.33
CS_2	5.15	2.04	16.82
F_2	2.84	1.42	4.49
H_2	1.48	0.74	0.64
HBr	2.84	2.28	11.59
HCl	2.54	1.98	7.82

HCN	3.25	1.50	5.74
HF	1.91	1.42	3.02
HI	3.22	2.66	17.89
I_2	5.32	2.66	29.56
N_2	2.60	1.50	4.59
O_2	2.67	1.46	4.47
V-shape molecules			*cylinder*
H_2O	2.27	1.69	5.12
H_2S	2.86	2.21	10.97
NO_2	3.33	2.20	12.67
O_3	3.48	2.24	13.75
SO_2	3.75	2.63	20.50
Planar trigonal molecules			*thin-prism*
BF_3	2.78	0.79	10.43
SO_3	3.94	2.04	13.69
Trigonal pyramidal molecules			*pyramid*
AsH_3	3.19	1.68	2.36
NH_3	2.38	1.49	1.22
PH_3	3.03	1.64	2.17
Tetrahedral molecules			*pyramid*
CBr_4	5.13	4.92	18.70
CCl_4	4.87	4.40	15.07
CF_4	351	3.14	5.56
CH_4	2.49	2.20	1.96
CI_4	6.15	5.59	30.54

It appears that many of these molecules (exceptions are CS_2, HI, I_2, O_3, SO_2, SO_3, CBr_4, CCl_4 and CI_4) may very well fit inside the cage of C_{60} molecule or to that matter, any species with a volume less than about 13 $A^{0\,3}$ may be able to fit inside the cage.

(c) **Diffusion Properties.** According to Stoke's law, the translational (f_{tra}) and the rotational (f_{rot}) frictional coefficients for a sphere with a radius r are given (53,54) by,

$$f_{tra} = 6 \pi \eta \, r \tag{2}$$

$$\text{and} \quad f_{rot} = 8 \pi \eta \, r^3 \tag{3}$$

where η is the coefficient of viscosity of the surrounding fluid. The above equations may be applied to C_{60} molecule provided that this molecule retains its spherical nature while undergoing rotational and transnational motions. The transnational diffusion (D_{tra}) and rotational diffusion (D_{rot}) coefficients are further related to their respective frictional coefficients as follows:

$$D_{tra} = kT / f_{tra} = k \, T / 6 \pi \eta \, r \tag{4}$$

$$\text{and} \quad D_{rot} = kT / f_{rot} = k \, T / 8 \pi \eta \, r^3 \tag{5}$$

where k is the Boltzmann constant and T is the Kelvin temperature. Relaxation time (τ_{rot}) applies only to rotational motion of the molecule and is related to the D_{rot} by Debye expression (55):

$$\tau_{rot} = 1 / 2 \, D_{rot} = 4 \pi \eta \, r^3 / kT \tag{6}$$

Thus, the diffusion coefficients and the relaxation time may be calculated from the above equations by knowing the radius and the viscosity. So, we have calculated $D_{tra} = 6.0 \times 10^{-6}$ cm^2/s, $D_{rot} = 3.62 \times 10^9$s^{-1} and $\tau_{rot} = 138$ ps for r =3.5 A^0 assuming $\eta = 1$cp, and T = 283 K. Since rotations around all the three axes are equal to one another for a sphere, the calculated D_{rot} and τ_{rot} values equally apply to all the three directions.

Johnson et al (22,23) have reported $\tau_{rot} = 9.2$ ps, 3.1 ps, and 15.5 ps for solid phase, gas phase , and in 1,1,2,2-tetrachloroethane indicating greater rotation in gas phase compared to either solid phase or in 1,1,2,2,-tetrachloroethane. Using these values, we have estimated the η as 0.20 cp, 0.067 cp, and 0.34 cp respectively for the solid phase, gaseous phase, and 1,1,2,2-tetrachloroethane for r =3.5A^0 (this value seems to be the approximate average radius of C$_{60}$ molecule indicated by many experiments, see Table II) and at T=283 K (this is the temperature at which these authors have reported their relaxation times). The calculated viscosities were further used in understanding the effect of the radius of C$_{60}$ molecule on diffusion coefficients and relaxations times (Table IV, Figures 5, 6, and 7).

Table IV. Translational and Rotational diffusion coefficients and Relaxation times for three viscosities at 283 K for C$_{60}$ molecule as a function of radius.

Radius(A^0)	D_{tra} (cm^2 s^{-1})			D_{rot}(s^{-1})			τ_{rot}(ps)		
	η(cp)			η (cp)			η(cp)		
	0.067	0.20	0.34	0.067	0.02	0.34	0.067	0.20	0.34
	x10^{-5}	x10^{-5}	x10^{-5}	x10^{-10}	x10^{-10}	x10^{-10}			
3.29	9.34	3.15	1.87	6.47	2.18	1.29	7.72	22.92	38.62
3.34	9.20	3.10	1.84	6.18	2.08	1.24	8.08	23.98	40.41
3.46	8.88	2.99	1.78	5.56	1.88	1.11	8.95	26,66	44.92
3.47	8.85	2.98	1.77	5.52	1.86	1.10	9.06	26.89	45.31
3.5	8.78	2.96	1.76	5.38	1.81	1.07	9.29	27.60	46.50
3.55	8.66	2.92	1.73	5.15	1.74	1.03	9.70	28.79	48.52
3.59	8.56	2.88	1.71	4.98	1.68	0.99	10.03	29.78	50.18
3.72	8.26	2.78	1.65	4.47	1.51	0.89	11.17	33.14	55.83
3.82	8.05	2.71	1.61	4.13	1.39	0.83	12.09	35.88	60.46

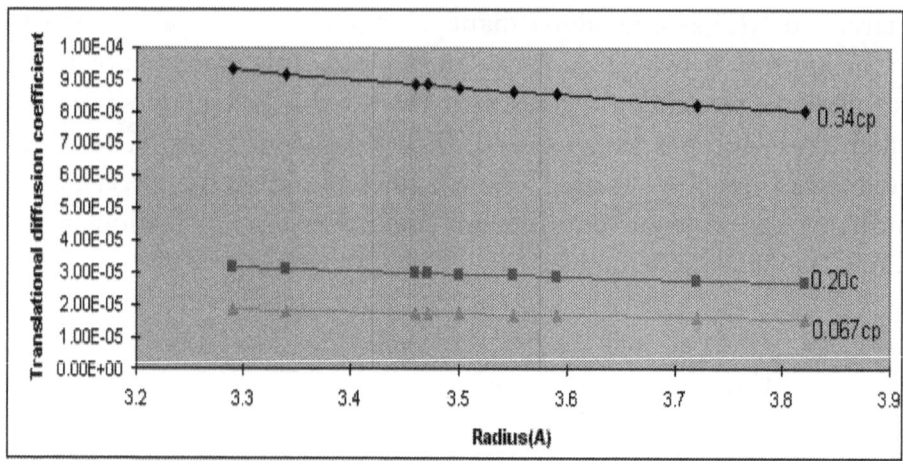

Figure 5. Plot of translational diffusion coefficient against the radius for three viscosity values shown on the graph(see also the Table IV).

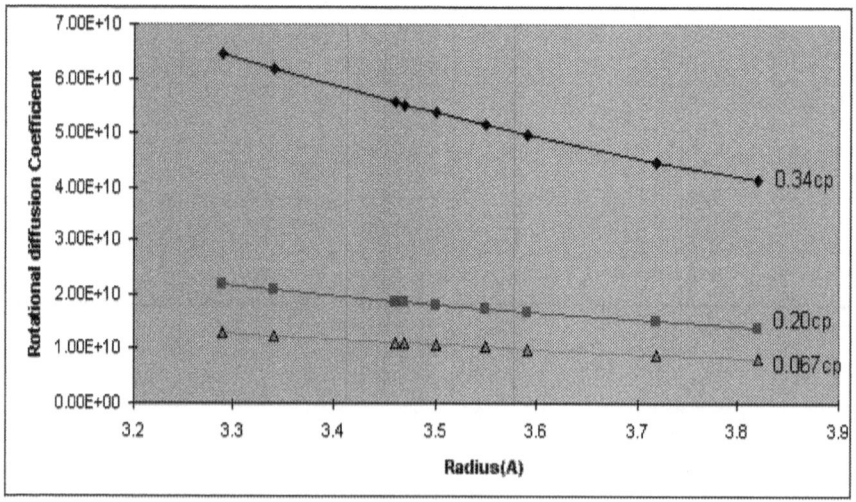

Figure 6. Plot of rotational diffusion vs. the radius for three viscosity values shown on the graph(see also Table IV).

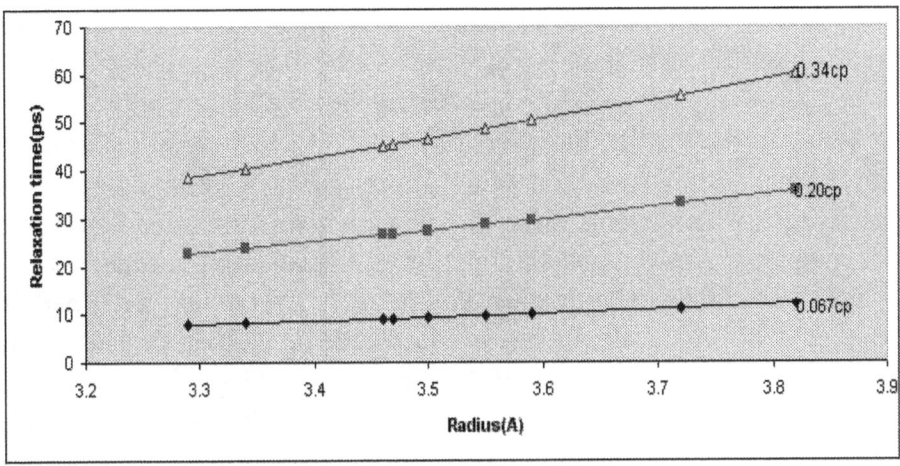

Figure 7. Plot of relaxation time vs. the radius for three viscosity values shown on the graph (see also Table IV).

It is clear that transnational and rotational diffusion coefficients decrease smoothly as the radius of the molecule increases. However, the decrease is more pronounced at higher viscosity (0.34 cp) than at lower viscosity in accordance with the common knowledge. On the other hand, the relaxation time increases as the radius increases. However, the increase in relaxation time, like transnational and rotational diffusion coefficient, is also appreciable at higher viscosity than at lower viscosity.

Since the buckyball is a hollow cage with windows created by pentagons and hexagons, many gaseous substances may very well diffuse freely into the cage. To understand this phenomenon, we need to calculate the size of the holes encircled by the pentagon and hexagon. The radius of inscribed circle is defined by R_{pen} = 0.68819 x L and R_{hex} = 0.86602 x L respectively for pentagon and hexagon (10,56). We have determined the diameter of 1.92 A^0 and 2.42 A^0 respectively for pentagon and hexagon using L = 1.395A^0 (standard aromatic bond length). Dimensions of some of the molecules are also listed in Table III. Therefore, it appears that any molecule with the diameter less than about 1.92 A^0 may be able to diffuse freely into the buckyball either through pentagons or hexagons. On the other hand, a molecule with diameter in the range of 1.92 - 2.42 A^0 may be able to pass only through hexagons. In reality, the diameters of pentagon and hexagon should be slightly smaller than those computed above due to pi electron density in these rings. In any event, these dimensions give some idea about the nature and the type of molecules that may diffuse into the cage.

(d) **Polarizability and Ionization Potential.** Polarizability is defined as the ease with which the electron cloud can be distorted. This is an important property in view of the fact that C_{60} molecule, being neutral, known to react with various ions inside and outside the cage, and also with number of molecules of different natures. Since C_{60} molecule being spherical in nature with no net dipole moment, it might be reacting through London's dispersion forces as a result of an induced dipole moment. As induced dipole moment manifests itself in the polarizability, it is of quite interesting to examine this property. The bond polarizability, α_θ , of a chemical bond placed in an electric filed with an angle, θ, is given by (57-60),

$$\alpha_\theta = \alpha_{||} \cos^2 \theta + \alpha_\perp \sin^2 \theta \tag{7}$$

Averaging over all orientations leads to the average of the bond polarizability, α_{bond} , which is,

$$\alpha_{bond} = 1/3 \, (\alpha_{||} + 2 \, \alpha_\perp) \tag{8}$$

where $\alpha_{||}$ is the polarizability in the direction $||$ to the bond direction, and $\alpha_\perp$ is the polarizability in the direction $\perp$ to the bond. The molecular polarizability is assumed to be the sum of all the bond polarizabilities, that is,

$$\alpha_{60} \; = \; \underset{\text{all bonds}}{\Sigma \; \alpha_{\text{bond}}} \qquad = \underset{\text{all bonds}}{\Sigma} \quad 1/3 \, (\alpha_{\parallel} + 2 \, \alpha_{\perp}) \qquad\qquad (9)$$

Using the equation (9), we have calculated the molecular polarizability $\alpha_{60} = 9.63 \times 10^{-23}$ cm^3 by using the bond polarizabilities of C_{ar} - C_{ar} bond as $\alpha_{\parallel} = 22.5 \times 10^{-25}$ cm^3, and $\alpha_{\perp} = 4.8 \times 10^{-25}$ cm^3(60).

In order to understand the ionization potential first we need to evaluate the relationship between the polarizability and the ionization potential. To do so, we have developed an empirical relationship by plotting ionization energy against the polarizability of about 60 substances (61,62) (Figure 8).

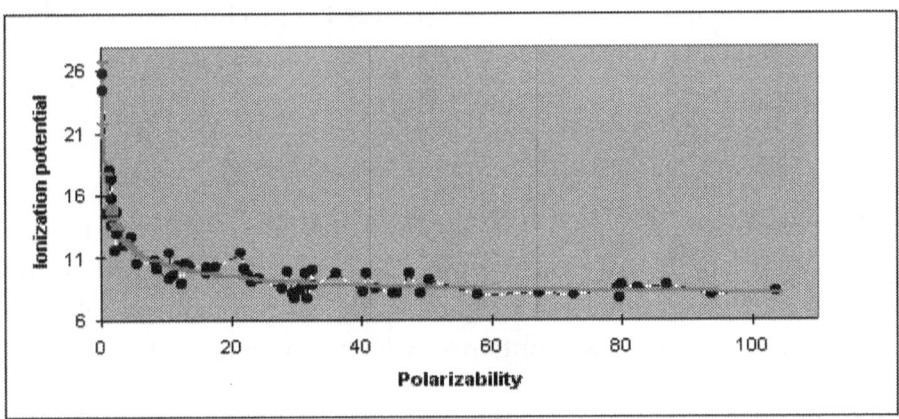

Figure 8. Plot of ionization potential as a function of polarizability for the molecules listed in references 61 and 62

The data may be fitted to either of the following equations:

$$I(eV) = a \cdot \exp \left[\, (\ln \alpha^{\cdot} - b \,)^2 / c \right] \qquad\qquad (10)$$

with $a = 7.926, b = 5.788,$ and $c = 44.78$ $(R^2 = 0.90)$

or $I (eV) = a + (b / \alpha^{*}) + (c / \alpha^{*2})$ (11)
    with $a = 8.844, b = 10.44, c = -1.492$ $(R^2 = 0.90)$

where $\alpha^{*} = \alpha \times 10^{24}$ cm^3. Small variance in the R^2 is due to the scatter in the experimental values. In any event, the smooth trend is evident in the plot(solid line in Figure 8). It is now possible to estimate the ionization potential for C_{60} molecule using either one of the above equations. The equation (10) and equation (11) respectively have yielded 8.2 eV and 8.9 eV both using $\alpha_{60} = 9.63 \times 10^{-23}$ cm^3 , which are in general agreement with the reported experimental values of 7- 8 eV (36,63).

The magnitude of the induced dipole moment may also be estimated using the above calculated polarizability. The induced dipole moment ($\mu_{induced}$) is proportional to the electric filed (E) as long as this field is not too strong, and is written as(59,60,61),

$$\mu_{induced} = \alpha_{60} E$$ (12)

We have estimated $\mu_{induced}$ ~3D for $E = 0.1$ V m^{-1}, ~29D for $E = 1.0$ Vm^{-1}, ~290D for $E = 10$ Vm^{-1} , and ~ 2888D for $E = 100$Vm^{-1}. When equation (12) was plotted as a function of E (Figure 9) a linear relationship with zero intercept was obtained, which is of course obvious from the above equation.

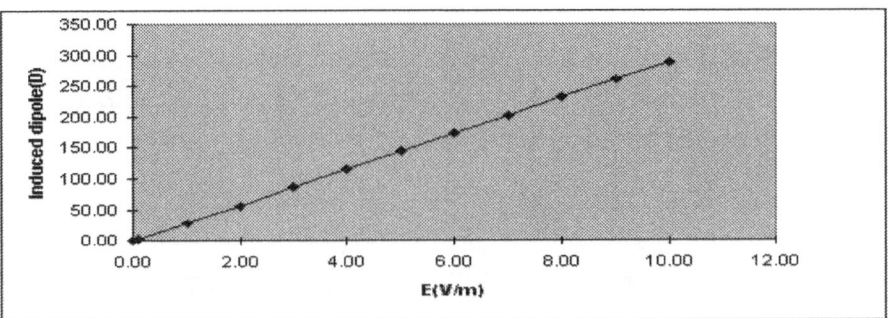

Figure 9. Plot of induced dipole moment computed by equation (12) as a function of electric field. The relationship is the straight line with zero intercept as indicated by the equation(12).

(e) Rotational Kinetic Energy. The relaxation time for C_{60} molecule in the gas phase is about 3 times faster than the solid phase, and 5 times faster than in liquid phase(in 1,1,2,2,-tetrachloroethane)(22,23). This in essence suggests that this molecule considerably executes the rotational motion in the gas phase. Hence, it would be of some interest to examine its rotational kinetic energy, which is given by the formula (64),

$$E_{kinetic} = 1/2 \ I \ \omega^2 \qquad (13)$$

where I is the moment of inertia, and ω is the angular velocity. For an object rotating through an angle θ (rad) in time t(s), its angular velocity is $\omega = \theta / t$ (rad/s). The moment of inertia I is related to the radius of gyration, that is, $I = m \ R_G^2$, where m is the total mass of the rotating body, and R_G is the radius of gyration. We have calculated $R_G = 3.5 \ A^0$

for C_{60} molecule, which is the same as the calculated radius. This is understandable especially in view of rigid spherical nature of this molecule. The dependency of the kinetic energy on the angular velocity was shown in Figure 10 for the radii given in the Table I.

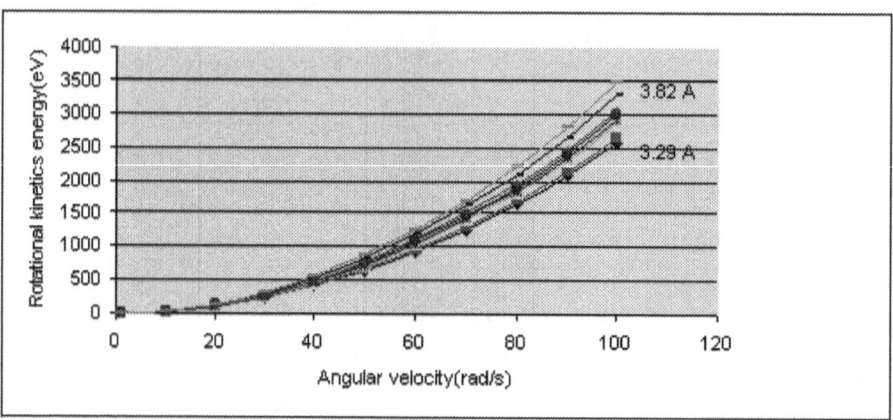

Figure 10. Plot of rotational kinetic energy vs. angular velocity for various radius of gyrations (3.29 A^0, 3.34 A^0, 3.46 A^0, 3.47 A^0, 3.5 A^0, 3.55 A^0, 3.59 A^0, 3.72 A^0, and 3.82 A^0)

Rotational kinetic energy increases exponentially as the angular velocity increases. The effect of radius appears to be more dominant at higher ω than at lower ω.

(f) Enthalpy of Formation. The enthalpy of formation (ΔH_f) of C_{60} molecule may be deduced as the sum of enthalpy of atomization(ΔH_{atom}) and enthalpy of bond dissociation energy(ΔH_{bde}). These two processes are:

$$60\ C\ (graphite)\ \rightarrow\ 60\ C \qquad \Delta H_{atom} = -27,450.0\ kJ\ mol^{-1} \qquad (14)$$

$$and\ \ 60\ C\ \rightarrow\ C_{60} \qquad \Delta H_{bde} = 42,840.0\ kJ\ mol^{-1} \qquad (15)$$

Then $\Delta H_f = \Delta H_{atom} + \Delta H_{bde} = -27,450.0 + 42,840.0 = 15,390$ kJ mol^{-1} = 3,678 kcal mol^{-1} In arriving this value, we have used 714 kJmol^{-1} for atomization enthalpy, and 330 kJmol^{-1} for single bond dissociation enthalpy, and 585 kJmol^{-1} for double bond dissociation enthalpy(65).

Huckel Molecular Orbital Method

(a) C - C Bond Length: We have carried out the Huckel molecular orbital calculations to investigate the nature of the bonds in C_{60} molecule. Like other experimental and theoretical results (see Table I), we also have encountered two types of bonds, namely, those between pentagons and hexagons (C_5 - C_6) and between hexagons and hexagons (C_6 - C_6). Huckel method does not give the bond length directly rather it gives the bond order(p_{rs}). Our calculated bond orders were 0.476 and 0.601 respectively for C_5 - C_6 and C_6 - C_6 . The bond orders may be converted into bond lengths either through the equation 16(65) or the equations 17 (66):

$$R(A^\circ) = 1.54 - \left(\frac{1.54 - 1.32}{\{1 + 0.765\,[\,(2 - P)/(P-1)]\,\}} \right) \tag{16}$$

where, $P = 1 + p_{rs}$, is the total bond order, and

$$R(A^\circ) = 1.50 - 0.16\,p_{rs} \tag{17}$$

Thus the computed bond lengths were C_5 - C_6 = 1.421 A$^\circ$ from equation(16) and 1.424 A^0 from equation(17) with an average of 1.42 A^0, and C_6 - C_6 = 1.394 A$^\circ$ from equation(16) and 1.404A$^\circ$ from equation(17) with an average of $1.39A^0$, which are in agreement with other experimental as well as theoretical results (Table I).

We have also investigated the effect of negative charges (C_{60}^{-n}, n=1,10) on the bond length, the radius, and the volume of the buckyball. In doing so, the bond lengths were first computed according to the equation(17) and the radii were estimated according to the equation(1) using the average bond length of C_5 - C_6 and C_6 - C_6. These radii values were further utilized in determining the volumes (spherical shape is assumed for C_{60} molecule). In order to understand the effect of the negative charge on these parameters, the percent expansions (= [Property of C_{60}^{-n} - Property of C_{60}] / [Property of C_{60} * 100), rather than their actual values, were plotted as a function of charges in Figure11.

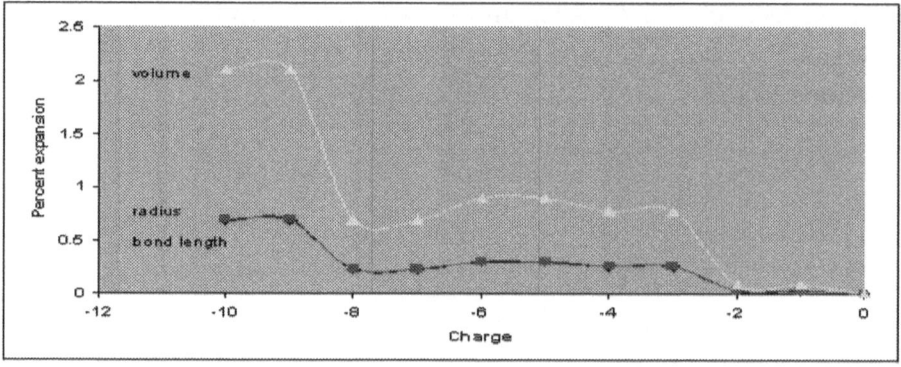

Figure 11. Plot of percent expansion in bond length, radius, and volume as a function of charge on C_{60} molecule.

It is clear that the negative charge has small but definite effect on the bond length - as the negative charge on the buckyball increases the bond length also increases slightly, the increase being greater for C_{60}^{-9} and C_{60}^{-10}. It is not clear, however, that whether this effect is real or an artifact of the Huckel calculation. In any event, a further study is needed to draw any definite conclusions regarding the influence of negative charge on the bond length.

(b) Net Pi (π) Charges: The net pi charges not only fluctuate from one carbon atom to another but also vary from one anion to another (Figure 12).

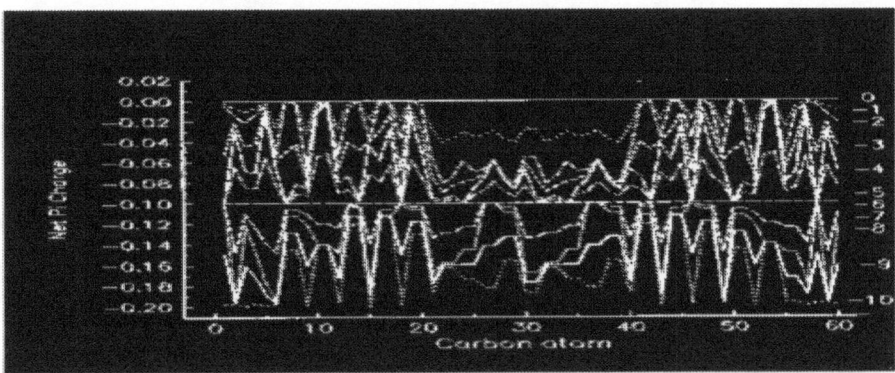

Figure 12. Variation in net pi charge(ordinate) on each carbon atom(abscissa) for various charges indicated on the graph.

Our computed charges indicate that these charge remain constant for carbon atoms only in C_{60} and C_{60}^{-6} species, probably indicating the chemical equivalency of all the atoms. Such uniform distribution of charges, of course, is understandable in a neutral molecule but it is quite puzzling to see in an anion carrying only -6 charge. When C_{60} molecule is reduced to C_{60}^{-1} , the non-uniform charge distribution is introduced. A further reduction in charge appears to have retained overall charge distribution symmetry(C_{60}^{-6} is an exception) even though the magnitudes of charges differ from one anionic form to another. The charge distribution may be approximately grouped into three distinct regions, namely, C_1- C_{21} (region I), C_{22}-C_{40} (region II), and C_{41}-C_{60}(region III)(Figure 12) neglecting the small fluctuations within the same region. This might be an indication that all the carbon atoms in anions do not react the same way; the atoms within the same region might behave similarly, but atoms in different regions might behave differently especially in a situation where the charge distribution plays an important role.

(c) **Delocalization Energy:** The calculated HOMO and LUMO energies were $\alpha + 0.618\beta$ and $\alpha - 0.139\beta$ respectively. The difference in theses energies, ΔE, was about 0.757β. The delocalization energy was computed to be about $31.168\ \beta$ or 0.519 per atom. The exchange integral β has been assigned numerous values in the literature (65,66). We have used 20 kcal/mol for β and have determined $\Delta E = 15.140$ kcal/mol and delocalization energy of 623.34 kcal/mol or 10.38 kcal/ (mol atom). The delocalization energy of benzene is about $2\ \beta$ or 0.33β / atom. Comparing this value with that of C_{60} molecule suggests that the delocalization energy for C_{60} molecule is about 57% more than benzene.

(d) **Superdelocalizability:** Superdeocalizability (Fukui) is the parameter often used to indicate the reactivity index. Our calculation showed that all the atoms in C_{60} molecule have the same value of 0.83 indicating that each atom is equally likely to participate in any kind of reaction. For anions investigated here, this index varies considerably: 0.34 to 0.84 for C_{60}^{-1} and C_{60}^{-2} ', 0.11 to 0.82 for C_{60}^{-3} and C_{60}^{-4} ', 0.11 for C_{60}^{-5} and C_{60}^{-6} ; -0.14 to -0.11 for C_{60}^{-7} and C_{60}^{-8} ; and -0.15 to -0.07 for C_{60}^{-9} and C_{60}^{-10}. This suggests that reactivity of carbon atoms in anions might be quite different from those in neutral molecule which is also suggested by our calculation on the net pi charge distribution.

(e) **Dipole Moment:** The dipole moments of various anionic forms of C_{60} molecule were computed and plotted in Figure 13, where a linear trend is exhibited; the dipole moment (μ) increases linearly as the negative charge increases.

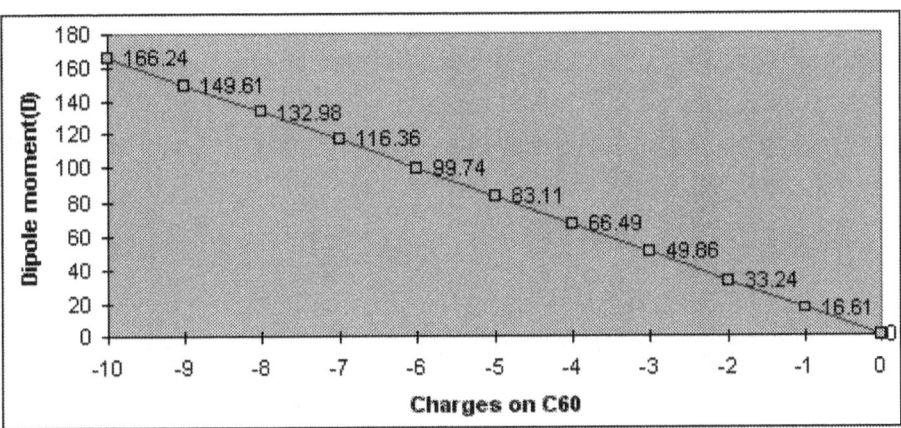

Figure 13. Plot of dipole moment (D) against negative charge on C_{60} molecule.

Wilson *et al* (67) have tabulated the half-wave potentials($E_{1/2}$) for number of anions in various systems. In Figure14 we have plotted these potentials against our calculated dipole moments.

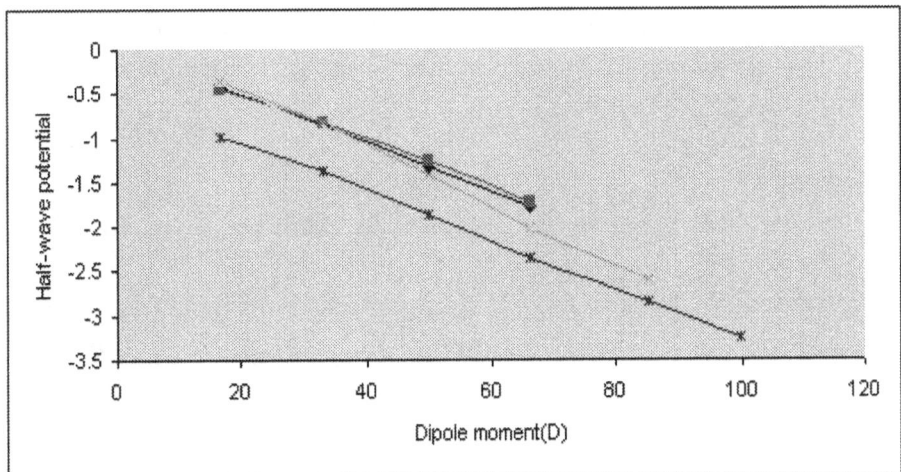

Figure 14. Plot of half-wave reduction potential($E_{1/2}$) against dipole moment(D) for various systems. The $E_{1/2}$ are taken from the ref.67. t

benzonitrile/TPAClO$_4$,RT, z CH$_2$Cl$_2$/TBABF$_4$, RT, Δ THF/TBAClO$_4$,RT, x benzene/THAClO$_4$, RT, and R (CH$_3$CN/toluene)/TBAPF$_6$, -10^{0}C .

There is an excellent correlation between our calculated dipole moments and their half-reduction potentials. These data were fitted to following least square equations:

$$E_{1/2} = 0.0346 - 0.272 . \mu (D), \quad \text{for (benzonitrile/TPAClO}_4, \text{RT)} \quad (R^2 = 1)$$

$$E_{1/2} = 0.0096 - 0.0257 . \mu (D), \quad \text{for (CH}_2\text{Cl}_2/\text{TBABF}_6, \text{RT)} \quad (R^2 = 1)$$

$$E_{1/2} = 0.2045 - 0.0330 . \mu (D), \quad \text{for (THF/TBACLO}_4, \text{RT)} \quad (R^2 = 1) \quad (18)$$

$$E_{1/2} = 0.2269 - 0.0332 . \mu (D), \quad \text{for (benzene/TBACLO}_4, \text{RT)} \quad (R^2 = 1)$$

$$E_{1/2} = -0.4892 - 0.0278 . \mu (D), \quad \text{for (CH}_3\text{CN/toluene/TBAPF}_6, -10^0\text{C)} \quad (R^2 = 1)$$

These equations may be of some useful in predicting E$_{1/2}$ of higher anions by knowing their dipole moments or vice versa.

Acknowledgment: The graphics in Figures 1 and 2 are done with the aid of RasMol. I sincerely thank Roger Sayle, Bimolecular Structure Department, Glaxo Wellcome Research Development, Stevenage, Hertforshire, UK, for providing this graphics tool.

References

1.   Fuller B.R., *Synergetics 2*, McMillan Publishing Co., Inc., New York, 1979.

2.   Kenner H., *Bucky - A Guided Tour of Buckminister Fuller*, William Morrow & Co.,Inc., New York, 1973. (b) Hatch A., *Buckminister Fuller - At Home in the Universe*, Crown Publishers, Inc, New York, 1974.

3.   Jacobson R.H., *Searching for the Soccer Ball : An unsuccessful Synthesis of 3a.Corannulene*. Ph.D. Dissertation, UCLA, 1986. (b) Xiong Y., *Synthesis of [1,1,1]Paracyclopane and Other Strained*

Aromatic Compounds, in Search For the Soccer Ball C_{60} ,Ph.D. Dissertation, UCLA, 1987. (c) Loguercio D., Jr., *Studies Toward a Convergent Synthesis of* C_{60} . Ph. D. Dissertation, UCLA, 1988. (d) Shen D., 3e. *Approaches to Soccherene (I_h - C_{60}) and Other Carbon Spheres,* Ph.D. Dissertation,UCLA, 1990.

4. Hoffman R., *Sci. Am.* 1993, 268, 66.

5. Kratschmer W., Lamb L.D., Fostiropoulos, and Huffman D.R., *Nature,* 1990, 347, 354.

6. Curl R.F. and Smalley R.E. *Science,* 1988, 242, 1017.

7. Kroto H.W., *Science,* 1988, 242, 1139.

8. Wenninger M.J., *Polyhedron Models,* Cambridge university Press, New York, 1971.

9. Holden A., *Shapes, Space and Symmetry,* Dover Publications, Inc., 1991.

10. Coxeter, H.S.M., *Regular Polytropes,* Dover Publications, Inc., 1973.

11. *C & E News,* American Chemical Society, Washington, D.C., December 23, 1991; December 16, 1991; September 16, 1991; November 9, 1992, 26; September 28, 1992; September 28, 1992; December 7,1992; November 2, 1992; July 13, 1992; June 1, 1992; January 4, 1993; January 25, 1993; January 28, 1993; March 8, 1993; June 21, 1993; October 25, 1993; January 3,1994; January 31,1994; February 7, 1994; February 14, 1994; February 28, 1994; *Popular Science,* August 1991, 52; Billups W.E. and Ciufolini M.A.Eds., *Buckminsterfullerenes,* VCH Publishers, Inc., N.Y. 1993.

12. Baseck P.R., Tsipursky S.J. and Hettick R., *Science (Washington, D.C.),* 1992, 257,215 ; *C & E News,* American Chemical Society, Washingon, D.C., July 13, 1992.

13. Fleming R.M., Hessen B., Siegrist T., Kortan A. R., Marsh P., Tycko R., Dabbagh G. and Hadden R.C., in *Fullerenes,* ACS Symposium

Series 481, Hammond G.S. and Kluck V.J., Eds., American Chemical Society, Washington, D.C., 1992, chapter 2.

14.   Gorum S.M., Greaney M.A., Day V.W., Day C.S., Upton R.M. and Briant C.E., in *Fullerenes* , ACS Symposium Series 481, Hammond G.S. and Kluck V.J., Eds., American Chemical Society, Washington, D.C., 1992, chapter 3.

15.   Fischer J.E., Heiney P.A., Luzzi D.E. and Cox D.E., in *Fullerenes*, ACS Symposium Series 481, Hammond G.S.and Kluck V.J., Eds., American Chemical Society, Washington,D.C., 1992, chapter 4.

16.   Hawkins J.M., Meyer A., Lewis T.A. and Loren, S., in *Fullerenes*, ACS Symposium Series 481, Hammond G.S.and Kluck V.J. Eds., American Chemical Society, Washington, D.C., 1992, chapter 6.; Hawkins J.M., Meyer A., Lewis T.A., Loren S. and Hollander F.S., *Science* (Washington, D.C.), 1991, 252, 312.

17.   Smalley R.E., in *Fullerenes*, ACS Symposium Series 481, Hammond G.S. and Kluck V.J. Eds., American Chemical Society, Washington, D.C., 1992, chapter 10.

18.   Fischer J.E., Heinley P.A. and Smith III A.B., *Acc. Chem. Res.*, 1992, 25, 112.

19.   Fagan P.J., Calabrese J.C. and Malone B., *Acc. Chem. Res.* ,1992, 25, 134 ; Fagen P.J., Calabrese J.C. and Malone B., *Science* (Washington, D.C.), 1991, 252,1160.

20.   Hawkins J.M., *Acc. Chem. Res.* ,1992, 25, 150.

21.   Cox D.M., Behal K., Creegan K., Disko M., Hsu C.S., Kollin E., Miler J., Robbins J., Robbins W., Sherwood R.D., Tindall P., Fisher D. and Meitzer G.,*Proc. Mater. Res.Soc.*, Boston, 1991, 206,651.

22.   Johnson R.D., Yannoni C.S., Salem J.R., Meijer G. and Bethune D.S., in *Fullerenes*, ACS Symposium Series 481, Hammond G.S.and Kluck V.J. Eds., American Chemical Society, Washington, D.C., 1992,chapter 7.

23.   Johnson R.D., Bethune D.S.and Yannoni C.S., *Acc. Chem. Res.*, 1992, 25, 169.

24. Yannoni C.S., Bernier P.P., Bethune D.S., Meijer G. and Salem J.R., *J. Am. Chem.Soc.*, 1991, 113, 3190.

25. Hadden R.C., Hebard A.F., Rosseinskey M.J., Murphy D.W., Glarum S.H., Palstra,T.T.M., Ramirez A.P., Duclos S.J., Fleming R.M., Siegrist, T. and Tycko, R., in *Fullerenes*, ACS Symposium Series 481, Hammond G.S.and KluckV.J. Eds., American Chemical Society, Washington, D.C., 1992, chapter 5.

26. Cox D.M., Sherwood R.D., Tindall P., Creegan K.M., Anderson W.and Martell D.,.in *Fullerenes*, ACS Symposium Series 481, Hammond G.S. and Kluck V.J. Eds., American Chemical Society, Washington, D.C.,1992, chapter 8.

27. Malhotra R., Lorents D.C., Bae Y.K., Becker C.H.,Tse D.S., Jusinsk L .E. and Wachsman E.D.,in *Fullerenes*, ACS Symposium Series 481, Hammond G.S. and Kluck V.J. Eds., American Chemical Society, Washington, D.C., 1992, chapter 9.

28. Wudl F., Hirsch A., Khenmani K.C., Suzuki T., Allemand P.M., Koch A., Eckert,H., Srdanov G., Webb H.M., in *Fullerenes*, ACS Symposium Series 481, HammondG.S. and Kluck V.J. Eds., American Chemical Society, Washington, D.C., 1992,chapter 12.

29. Fagan P.J., Calabrese J.C. and Malone B., in *Fullerenes*, ACS Symposium Series 481, Hammond G.S. and Kluck V.J., Eds., American Chemical Society, Washington, D.C.,1992,chapter 11.

30. Hadden R.C., *Acc. Chem. Res.*, 1992, 25, 127.

31. Smalley R.E., *Acc. Chem. Res.*, 1992,25, 98.

32. Heath J.R., in *Fullerene*, ACS Symposium Series 481, Hammond G.S. and Kluck V.J., Eds., American Chemical Society, Washington, D.C., 1992, chapter 1.

33. Whetten R.L. and Diederich F., *Acc. Chem. Res*, 1992, 25, 119.

34. Weaver J.H., *Acc. Chem. Res.*, 1992, 25, 143.

35. Wudl F., *Acc. Chem. Res.*, 1992, 25, 157.

36. McElvancy S.W., Ross M.M. and Callahan J.H., *Acc. Chem. Res.*, 1992, 25, 162.

37. Hebard A.I., Rossenskey M.J., Hadden R.C., Murphy D.W.,Glarum S.H.,Palstra, Ramire T.T.M.and Kortran, A.P., *Nature, A.R.* (London), 1991, 350, 600.

38. Stephens P.W., Mihaley L., Lee, P.L., Whetten R.L., Huang S.M., Kaner P., Diederich, F. and Holczer K. *Nature*(London), 1991, 351, 632.

39. Barth W.E. and Lawton R.G., *J.Am. Chem.Soc.*, 1966, 88, 380.; Barth, W.E. and Lawton R.G., *J. Am. Chem. Soc.* 1971, 93, 1730.

40. Scott L.T., Hashem M.M. and Bratcher M.S., *J. Am. Chem. Soc.*,1992, 114, 1920-1921 ; Borchardt A., Fuchiello A., Kilway K.V., Baldridge K.K. and Siegel J.S., *J. Am. Chem. Soc.* 1992, 114, 1921-1923.

41 Scott L.T., Hashemi M.M., Meyer, D.T. and Warren H.B., *J. Am. Chem. Soc.*, 1991, 113, 7092.

42. *C & E News*, American Chemical Society, 993, January 25, 20.

43. Wetner W., Jr., and Van Zee R.J., *Chem. Rev.*, 1989, 89, 1713.

44. Scuseria G.E., *Chem. Phys. Lett.*, 1991, 176, 423.; Scaseria G.E.,in *Buckminsterfullerenes*, Billups W.E. and Ciufolini, M.A., Eds., VCH Publishers, New York 1993, chapter 5.

45. Dias J.T., *Chem. Edu.*, 1989, 66,1012.

46. Amic D. and Trinajastic N.J. *Che. Soc. Perkin Trnas.*, 2 1990, 1595.

47. Newton M.D. and Standon R.E., *J.Am.Chem.Soc.*, 1986,108,2569.

48. Hare J.P. and Kroto H.W., *Acc. Chem. Res.*, 1992, 25, 106.

49. Pople J. A. and Beverdige D. L. *Approximate Molecular Orbital Theory,* McGraw-Hill Book Company, New York 1970.

50. Press W.H.,Flannery B.P.,Teukolsky,S.A. and Vetterling,W.J., *Numerical Recipes,The Art of Scientific Computing*, Cambridge University Press, N.Y. 1986, chapter 10.

51. Kumbar M., *J. Theor. Biol.*, 1975, 51, 307.

52. Kumbar M. and Siva Sankar D.V., *J.Am.Chem. Soc.*, 1975, 97, 7411.

53.   Tandford C, *Physical Chemistry of Macomolecules,* John Wiley & Sons, New York,1961.

54.   Kumbar M., *J. Macromol. Sci-Chem.,* 1983, A19,155.

55.   Edsall J.T., *Rotatory Brownian Movement* in Cohen E.E. and Edsall J.T., *Proteins, Aminiacids, and Peptides,* Reinhold Publishing Corp, N.Y. 1943, Chapter 22.

56.   Selby M.S. and Girling, B., *Standard Mathematical Tables,* The Chemical Rubber Co., Cleveland, Ohio 1965.

57.   Hirschfelder J.O., Curtiss C.F.and Bird, R.B., *Molecular Theory of Gases and Liquids,* Wiley, New York 1954.

58.   Flory P.J., *Statistical Mechanics of Chain Molecules,* Interscience Publishers, New York, 1969.

59.   Smith R.P.and Mortensen E.M., *J.Chem.Phys.,* 1960,32,502.

60   Volkenstein M.V., *Configurational Statistics of Polymeric Chains,* Interscience Publishers, New York 1963.

61.   Kauzman W. , *Quantum Chemictry,* Academic Press Inc., New York 1957.

62.   Moelwyn-Hughes E.A., *Physical Chemistry,* Pergamon Press, New York, 1961.

63.   Larson S.,Volosov A.and Rosen A., *Chem. Phys. Lett.,* 1987, 137,501.

64.   Tippens P.E., *Applied Physics,* McGraw-Hill Book Company, N.Y. 1978.

65.   Isaacs N.S., *Physical Organic Chemistry,* John Wiley and Sons, Inc., New York 1987.

66.   Murrell J.N., Ketle S.F.A. and Tedder, J.M., *Valance Theory,* John-Wiley and Sons, N.J.1969; Murrell, J.N., Kettle S.F.A., Tedder J.M., *The Chemical Bond,* John- Wiley and Sons, Inc., New York 1978.

67.   Wilson L.J., Flanagan S., Chibante L.P.F. and Alford J.M., in *Buckminsterfullerenes,* Billups W.E. and Ciufolini M.A., Eds., VCH Publishers, New York 1993, chapter 11.

APPENDIX B-1

Coordinates for Buckminsterfullerene (C_{60}). The following data structure is presented in alchemy file format to use with RASMOL. These coordinates can also be used with any other program.

60 atoms,   90 bonds, 0 charges, Buckminsterfullerene

19	C19	1.5698	-1.8281	-0.2204
20	C20	0.9055	-2.2598	-1.3711
21	C21	1.6351	-2.6944	-2.4797
22	C22	0.9710	-2.262	-3.6302
23	C23	1.7023	-1.8306	-4.7403
24	C24	1.2918	-0.7004	-5.4515
25	C25	0.1496	0.0005	-5.0533
26	C26	0.1498	1.3975	-5.0529
27	C27	1.2923	2.0969	-5.4509
28	C28	1.7029	3.2271	-4.7398
29	C29	0.9717	3.6589	-3.6298
30	C30	1.6363	4.0901	-2.4792
31	C31	3.0332	4.0909	-2.4366
32	C32	3.6975	3.6592	1.2860
33	C33	2.9655	3.2264	-0.1768
34	C34	3.3761	2.0959	0.5339
35	C35	4.5195	1.3970	0.1376
36	C36	4.5195	0.0000	0.1376
37	C37	3.3764	-0.6985	0.5343
38	C38	2.9661	-1.8283	-0.1776
39	C39	3.6982	-2.2608	-1.2870
40	C40	3.0339	-2.6925	-2.4376
41	C41	3.7637	-2.2592	-3.5448
42	C42	3.0994	-1.8275	-4.6954
43	C43	3.5524	-0.6979	-5.3813

44	C44	2.4357	0.0006	-5.8467
45	C45	2.4357	1.3976	-5.8467
46	C46	3.5525	2.0962	-5.3813
47	C47	3.1001	3.2264	-4.6961
48	C48	3.7644	3.6581	-3.5455
49	C49	4.8814	2.9597	-3.0800
50	C50	4.8399	2.9598	-1.6836
51	C51	5.2504	1.8296	-0.9725
52	C52	5.7026	0.6990	-1.6578
53	C53	5.2505	-0.4314	-0.9725
54	C54	4.8399	-1.5616	-1.6836
55	C55	4.8814	-1.5614	-3.0800
56	C56	5.3335	-0.4311	-3.7653
57	C57	4.6692	0.0006	-4.9159
58	C58	4.6692	1.3976	-4.9159
59	C59	5.3335	1.8293	-3.7653
60	C60	5.7441	0.6991	-3.0541
1	1	2	AROMATIC
2	1	5	AROMATIC
3	1	18	AROMATIC
4	2	3	AROMATIC
5	2	6	AROMATIC
6	3	4	AROMATIC
7	3	9	AROMATIC
8	4	5	AROMATIC
9	4	12	AROMATIC
10	5	15	AROMATIC
11	6	7	AROMATIC
12	6	20	AROMATIC
13	7	8	AROMATIC
14	7	22	AROMATIC
15	8	9	AROMATIC

16	8	25	AROMATIC
17	9	10	AROMATIC
18	10	11	AROMATIC
19	10	26	AROMATIC
20	11	12	AROMATIC
21	11	29	AROMATIC
22	12	13	AROMATIC
23	13	14	AROMATIC
24	13	30	AROMATIC
25	14	15	AROMATIC
26	14	33	AROMATIC
27	15	16	AROMATIC
28	16	17	AROMATIC
29	16	34	AROMATIC
30	17	18	AROMATIC
31	17	37	AROMATIC
32	18	19	AROMATIC
33	19	20	AROMATIC
34	19	38	AROMATIC
35	20	21	AROMATIC
36	21	22	AROMATIC
37	21	40	AROMATIC
38	22	23	AROMATIC
39	23	24	AROMATIC
40	23	42	AROMATIC
41	24	25	AROMATIC
42	24	44	AROMATIC
43	25	26	AROMATIC
44	26	27	AROMATIC
45	27	28	AROMATIC
46	27	45	AROMATIC
47	28	29	AROMATIC

48	28	47	AROMATIC
49	29	30	AROMATIC
50	30	31	AROMATIC
51	31	32	AROMATIC
52	31	48	AROMATIC
53	32	33	AROMATIC
54	32	50	AROMATIC
55	33	34	AROMATIC
56	34	35	AROMATIC
57	35	36	AROMATIC
58	35	51	AROMATIC
59	36	37	AROMATIC
60	36	53	AROMATIC
61	37	38	AROMATIC
62	38	39	AROMATIC
63	39	40	AROMATIC
64	39	54	AROMATIC
65	40	41	AROMATIC
66	41	42	AROMATIC
67	41	55	AROMATIC
68	42	43	AROMATIC
69	43	44	AROMATIC
70	43	57	AROMATIC
71	44	45	AROMATIC
72	45	46	AROMATIC
73	46	47	AROMATIC
74	46	58	AROMATIC
75	47	48	AROMATIC
76	48	49	AROMATIC
77	49	50	AROMATIC
78	49	59	AROMATIC
79	50	51	AROMATIC

80	51	52	AROMATIC
81	52	53	AROMATIC
82	52	60	AROMATIC
83	53	54	AROMATIC
84	54	55	AROMATIC
85	55	56	AROMATIC
86	56	57	AROMATIC
87	56	60	AROMATIC
88	57	58	AROMATIC
89	58	59	AROMATIC
90	59	60	AROMATIC

APPENDIX B-2

Coordinates for the structure in Figure 2 (IV) to use with RASMOL (alchemy file format). These coordinates can also be used to generate the structures in Figure 2(II) and (III).

90 atom, 90 bonds,   0 charges, Fullerene figure 2(Iv)

1	C1	0.0000	0.0000	0.0000
2	C2	1.3286	-0.4317	0.0000
3	C3	2.1498	0.6985	0.0000
4	C4	1.3286	1.8287	0.0000
5	C5	0.0000	1.3970	0.0000
6	C6	1.3286	-0.4317	0.0000
7	C7	1.8968	-1.7079	0.0000
8	C8	3.2862	-1.8539	0.0000
9	C9	4.1073	-0.7237	0.0000
10	C10	3.5391	0.5525	0.0000
11	C11	2.1498	0.6985	0.0000
12	C12	2.1498	0.6985	0.0000

13	C13	3.5391	0.8445	0.0000
14	C14	4.1073	2.1207	0.0000
15	C15	3.2862	3.2509	0.0000
16	C16	1.8968	3.1049	0.0000
17	C17	1.3286	1.8287	0.0000
18	C18	1.3286	1.8287	0.0000
19	C19	1.6191	3.1952	0.0000
20	C20	0.5809	4.1299	0.0000
21	C21	-0.7477	3.6982	0.0000
22	C22	-1.0382	2.3318	0.0000
23	C23	0.0000	1.3970	0.0000
24	C24	0.0000	1.3970	0.0000
25	C25	-1.2098	2.0955	0.0000
26	C26	-2.4197	1.3970	0.0000
27	C27	-2.4197	0.0000	0.0000
28	C28	-1.2098	-0.6985	0.0000
29	C29	0.0000	0.0000	0.0000
30	C30	0.0000	0.0000	0.0000
31	C31	-1.0382	-0.9348	0.0000
32	C32	-0.7477	-2.3012	0.0000
33	C33	0.5809	-2.7329	0.0000
34	C34	1.6191	-1.7982	0.0000
35	C35	1.3286	-0.4317	0.0000
36	C36	1.8968	-1.7079	0.0000
37	C37	1.3286	-2.9841	0.0000
38	C38	2.3668	-3.9189	0.0000
39	C39	3.5766	-3.2204	0.0000
40	C40	3.2862	-1.8539	0.0000
41	C41	3.5391	0.8445	0.0000
42	C42	4.5773	-0.0902	0.0000
43	C43	5.7871	0.6083	0.0000
44	C44	5.4967	1.9747	0.0000

45	C45	4.1073	2.1207	0.0000
46	C46	1.6191	3.1952	0.0000
47	C47	2.8289	3.8937	0.0000
48	C48	2.5385	5.2601	0.0000
49	C49	1.1491	5.4062	0.0000
50	C50	0.5809	4.1299	0.0000
51	C51	-1.2098	2.0955	0.0000
52	C52	-1.5003	3.4620	0.0000
53	C53	-2.8896	3.6080	0.0000
54	C54	-3.4578	2.3318	0.0000
55	C55	-2.4197	1.3970	0.0000
56	C56	-1.0382	-0.9348	0.0000
57	C57	-2.4275	-0.7887	0.0000
58	C58	-2.9957	-2.0650	0.0000
59	C59	-1.9576	-2.9997	0.0000
60	C60	-0.7477	-2.3012	0.0000
61	C61	3.2862	-1.8539	0.0000
62	C62	3.8544	-3.1302	0.0000
63	C63	5.2437	-3.2762	0.0000
64	C64	6.0649	-2.1460	0.0000
65	C65	5.4967	-0.8698	0.0000
66	C66	4.1073	-0.7237	0.0000
67	C67	4.1073	2.1207	0.0000
68	C68	5.4967	2.2668	0.0000
69	C69	6.0649	3.5430	0.0000
70	C70	5.2437	4.6732	0.0000
71	C71	3.8544	4.5272	0.0000
72	C72	3.2862	3.2509	0.0000
73	C73	0.5809	4.1299	0.0000
74	C74	0.8714	5.4964	0.0000
75	C75	-0.1668	6.4312	0.0000
76	C76	-1.4954	5.9995	0.0000

77	C77	-1.7859		4.6330	0.0000
78	C78	-0.7477		3.6982	0.0000
79	C79	-2.4197		1.3970	0.0000
80	C80	-3.6295		2.0955	0.0000
81	C81	-4.8394		1.3970	0.0000
82	C82	-4.8394		0.0000	0.0000
83	C83	-3.6295		-0.6985	0.0000
84	C84	-2.4197		0.0000	0.0000
85	C85	-0.7477		-2.3012	0.0000
86	C86	-1.7859		-3.2360	0.0000
87	C87	-1.4954		-4.6025	0.0000
88	C88	-0.1668		-5.0342	0.0000
89	C89	0.8714		-4.0994	0.0000
90	C90	0.5809		-2.7329	0.0000
1	2	1	AROMATIC		
2	3	2	AROMATIC		
3	4	3	AROMATIC		
4	5	4	AROMATIC		
5	5	1	AROMATIC		
6	6	1	AROMATIC		
7	7	6	AROMATIC		
8	8	7	AROMATIC		
9	9	8	AROMATIC		
10	10	9	AROMATIC		
11	11	10	AROMATIC		
12	12	2	AROMATIC		
13	13	12	AROMATIC		
14	14	13	AROMATIC		
15	15	14	AROMATIC		
16	16	15	AROMATIC		
17	17	16	AROMATIC		
18	18	12	AROMATIC		

19	19	18	AROMATIC	
20	20	19	AROMATIC	
21	21	20	AROMATIC	
22	22	21	AROMATIC	
23	23	22	AROMATIC	
24	24	1	AROMATIC	
25	25	24	AROMATIC	
26	26	25	AROMATIC	
27	27	26	AROMATIC	
28	28	27	AROMATIC	
29	29	28	AROMATIC	
30	30	23	AROMATIC	
31	31	30	AROMATIC	
32	32	31	AROMATIC	
33	33	32	AROMATIC	
34	34	33	AROMATIC	
35	35	34	AROMATIC	FIGURE 2(II)
36	36	6	AROMATIC	
37	37	36	AROMATIC	
38	38	37	AROMATIC	
39	39	38	AROMATIC	
40	40	39	AROMATIC	
41	41	12	AROMATIC	
42	42	41	AROMATIC	
43	43	42	AROMATIC	
44	44	43	AROMATIC	
45	45	44	AROMATIC	
46	46	18	AROMATIC	
47	47	46	AROMATIC	
48	48	47	AROMATIC	
49	49	48	AROMATIC	
50	50	49	AROMATIC	

51	51	23	AROMATIC
52	52	51	AROMATIC
53	53	52	AROMATIC
54	54	53	AROMATIC
55	55	54	AROMATIC
56	56	30	AROMATIC
57	57	56	AROMATIC
58	58	57	AROMATIC
59	59	58	AROMATIC
60	60	59	AROMATIC
61	61	7	AROMATIC
62	62	61	AROMATIC
63	63	62	AROMATIC
64	64	63	AROMATIC
65	65	64	AROMATIC
66	66	65	AROMATIC
67	67	13	AROMATIC
68	68	67	AROMATIC
69	69	68	AROMATIC
70	70	69	AROMATIC
71	71	70	AROMATIC
72	72	71	AROMATIC
73	73	19	AROMATIC
74	74	73	AROMATIC
75	75	74	AROMATIC
76	76	75	AROMATIC
77	77	76	AROMATIC
78	78	77	AROMATIC
79	79	25	AROMATIC
80	80	79	AROMATIC
81	81	80	AROMATIC
82	82	81	AROMATIC

FIGURE 2(III)

83	83	82	AROMATIC	
84	84	83	AROMATIC	
85	85	31	AROMATIC	
86	86	85	AROMATIC	
87	87	86	AROMATIC	
88	88	87	AROMATIC	
89	89	88	AROMATIC	
90	90	89	AROMATIC	FIGURE 2(IV)

Appendix C
Fullerene II. Computer Simulation of Clusters of C_{60} Molecules

Sujata Kumbar and Mahadev Kumbar (1994)

Buckministerfullerene (C_{60}) has been studied extensively using variety of methods including solid state X-ray , NMR, and quantum mechanical (1-9). Among the parameters that were investigated extensively, bond lengths, radius, and intermolecular separation (see Table I and II) are of considerable value due to their important role in understanding various physical and chemical properties of the C_{60} molecule.

Table I. Bond length (A^{0}) in C_{60} molecule from various sources.

Compound	C_5 - C_6	C_6 - C_6	Method	Reference
C_{60} - cyclohexane	1.51	1.44	X-ray	2
C_{60} - Osmylated	1.432	1.388	X-ray	3
C_{60} - Pt derivatives	1.45	1.39	X-ray	4
C_{60}	1.45	1.40	NMR	5
C_{60}	1.42	1.42	EXAFS	6
C_{60}	1.36-1.45	1.34-1.38	THEOR	7,8
C_{60}	1.42	1.394	Huckel	9

Table II. Shortest distance (molecular separation) between two C_{60} molecules.

Substance	Distance (A)	$\Delta^{\#}$	Method	Ref.
$K_6 C_{60}$	9.86	-0.06	X-ray	10,11
$Rb_6 C_{60}$	9.98	0.06	X-ray	10,11
$Cs_6 C_{60}$	10.21	0.29	X-ray	10,11
Pentane-solvated C_{60}	10.04	0.12	X-ray	12
Cyclohexane-C_{60}	12.218	2.298	X-ray	2
Osmylated -C_{60}	10.21	0.29	X-ray	13
C_{60}	10 -11	0.08-1.08	STM	14,15
C_{60} -undoped	10.02	0.10	X-ray	16,17,18
C_{60}	9.92	0.0	THEOR	This work

The difference between distances in column 2 and our calculated distance (9.92 A0)

Majority of the studied conducted were based on the solid state intercalated with either metal ions or neutral molecules. It is evident from these studies that the intercalated species do not influence the radius of C_{60} molecule to a greater extent. This is understandable, especially, in view of the fact that C_{60} molecule is a rigid spherical molecule with dimensions much greater than added species. Nevertheless, added species tend to influence bond lengths of pentagons and hexagons only in the immediate neighborhood of binding site(s), but have no effect whatsoever on those bond lengths that are far remote from the binding site(s). In addition, crystallographic studies(10-13) also have pointed out that the intermolecular separation (defined as the distance between the centers of C_{60} molecules), to some degree, depends upon the nature and size of the species incorporated into the C_{60} matrix. Hence, it is of some interest to understand this separation in absence of these species in order to assess their degree and nature of influence. Therefore, this study is undertaken to investigate intermolecular separation in absence of added species and further to explore its influence on the bulk density of this allotropic form. To accomplish this, we have generated dimer(cluster of two molecules $(C_{60})_2$) and trimer (cluster of three molecules $(C_{60})_3$) on the computer. The detail procedure for computer simulation is described below.

Computer Simulation

Each buckyball is generated according to the procedure described in part I. Then the clusters were simulated by transforming local coordinates of each C_{60} molecule to a global coordinate system. The optimum intermolecular separation is calculated by minimizing the energy of the cluster (E_{inter}^{x-mer}) containing two C_{60} molecules (dimer) and three C_{60} molecules (trimer). In computing the energy of the cluster, the pair-wise addition concept is assumed, that is,

$$E_{inter}^{x-mer} = \sum_{m}^{x-1} \sum_{n}^{x} E_{mn} \quad ; x = 2,3 \qquad (1)$$
$$(m < n)$$

where E_{mn} is the intermolecular interaction energy between mth and nth C_{60} molecules. This energy is calculated as the sum of the non-bonded energy (E_{nb}) and the dispersion energy(E_{disp}):

$$E_{mn} = E_{nb} + E_{disp} \qquad (2)$$

The non-bonded interaction energy between the pair of interacting atoms preset on two different C_{60} molecules is computed by using Lennard-Jones 6-12 potential function, which is,

$$E_{nb} = \sum_{i,j} d_{ij} r_{ij}^{-12} - e_{ij} r_{ij}^{-6} \qquad (3)$$

The first term in the above equation represents the repulsive forces and the second term describes the attractive forces between atom i and j present on two different molecules. The term r is the distance between the ith and jth interacting atoms, and d_{ij} and e_{ij} are the coefficients associated with the repulsive and attractive terms respectively, and are determined by the procedure described elsewhere[19,20]. The dispersion energy for a pair of identical interacting molecules is added through the following equation[21-23].

$$E_{disp} = -17.30 \times 10^{18} \ (\ I\alpha^2 \ / \ R^6 \ ) \ (\ kcal/mol) \tag{4}$$

where I is the ionization energy in eV, α is the polarizability in cm^3, and R is the distance between two molecules. In evaluating E_{disp} energy, we have taken the value of α calculated in part I. The I is determined using the equation(2) of part I, and R was set to the distance between the center of masses of two interacting molecules. Once the E_{inter}^{x-mer} is calculated with above described procedure, it was then minimized using the Powell minimization technique (24) to determine the minimum separation between two molecules (dimer) and three molecules(trimer).

Results and Discussion

(a) **Intermolecular Separation.** The separation in dimer and trimer respectively were found to be 3.15 A^0 (Figure 1) and 3.0 A^0 (Figure 2). A small decrease (0.15 A^0) in the separation from dimer to trimer may be due to an added attraction exerted by the presence of third molecule.

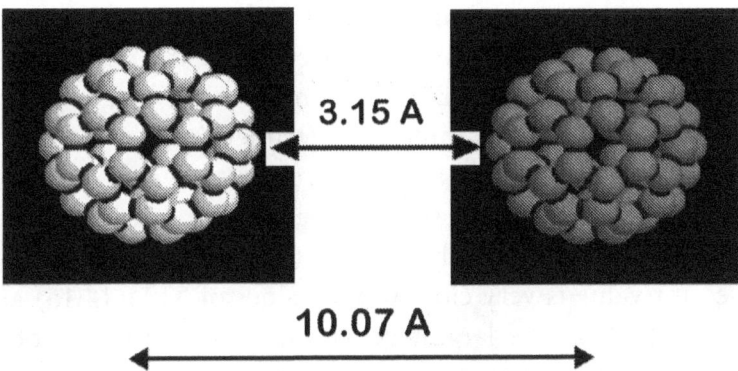

Figure 1. Computed values for molecular separation and intermolecular separation in dimmer.

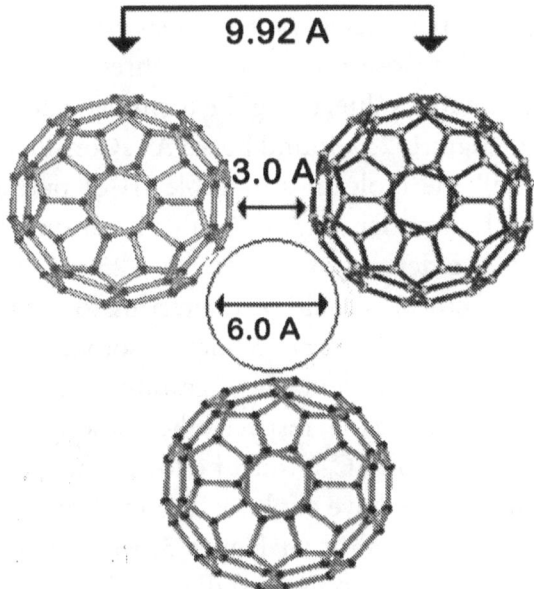

Figure 2. Computed values for molecular separation, intermolecular separation, and the size of the hole in trimer.

The intermolecular separations in dimer and trimer were estimated to be about 10.07 A^0 and 9.92 A^0 respectively using the radius of 3.46 A^0 for C_{60} molecule with standard C-C aromatic bond length of 1.39 A^0 (see part I). Since three-body interaction (trimer) is more realistic model than two-body interaction(dimer), the computed value of 9.92 A^0 for the intermolecular separation seems to be more reasonable. Comparing our value with values from other studies (Table II) few observation can be made: our value is very close to the values of STM[14,15] and un-doped C_{60} [16-18]. The largest deviation from our values occurs in cyclohexane-C_{60} [2]. Other experimental values, except $K_6 C_{60}$, also fall closer or above our value. It should be noted, however, that our value is based an ideal situation without any external influence of any kind - the difference might be attributed to size and nature of doped species.

In addition to this parameter, we have also estimated the radius of the hole created by the closest approach of three molecules in trimer (Figure 2). Our estimated value, using the formula, 0.28867 L, for equi-lateral triangle of length L(25), found be ~3 A^0 . Considering the size of this hole, almost all the molecules in Table III of part I may diffuse freely through this hole either longitudinally or transversely.

(b) **Bulk Density.** The density of graphite is 2.25 g / cm^3 and that of diamond is 3.514 g / cm^3. It should be interesting to predict the density of solid C_{60}. Fischer *et al* [11] have reported isothermal volume com-pressibility , $-1/V(dV/dP)$, for solid C_{60} , graphite, and diamond as 6.9, 2.7, and 0.18 x 10^{-12} cm^2 / dyne, respectively. These values clearly indi-cate that the density of solid C_{60} must be lower than graphite due its higher compressibility. We have explored the relationship between the density and the compressibility. In Figure 3, we have plotted density against compressibility for graphite and diamond.

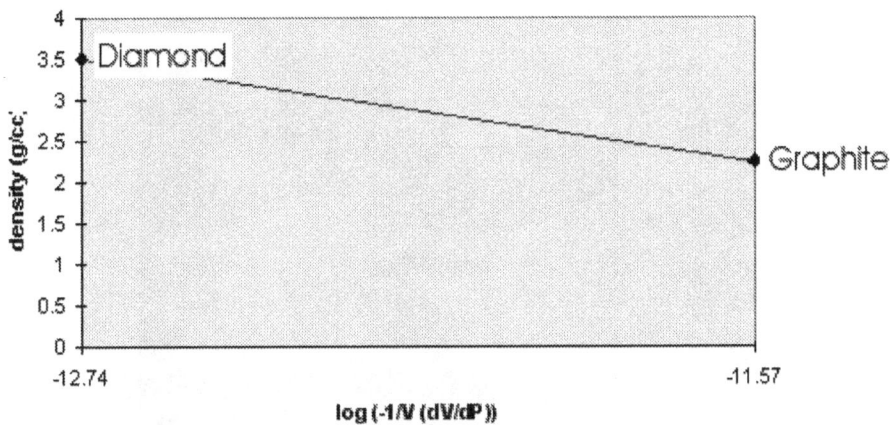

Figure 3. Plot of density (g/cc) for graphite and diamond as a function of compressibility.

Assuming the linear trend still holds when density of solid C_{60} is included, the data on graphite and diamond was fitted to the following equation:

$$\text{density} = -10.183 + (-1.074) \times [\log(-1/V\,(dP/dV))] \text{ (g/cc)} \qquad (5)$$

Using the compressibility of solid C_{60} (6.9×10^{-12} cm²/dyne), we have determined the density of this material as 1.81 g/cm³, which lies below the graphite as it should be by virtue of its compressibility data. However, one should be very cautious using this value as it is based on the assumption of linear relationship between the density and the compressibility, which may or may not hold when the true value of C_{60} is included. Nonetheless, it is tempting to explore further the relationship between the density and the intermolecular separation.

Let R be the molecular radius and r be the intermolecular separation (Figure 4) both in A^0 unit. Then,

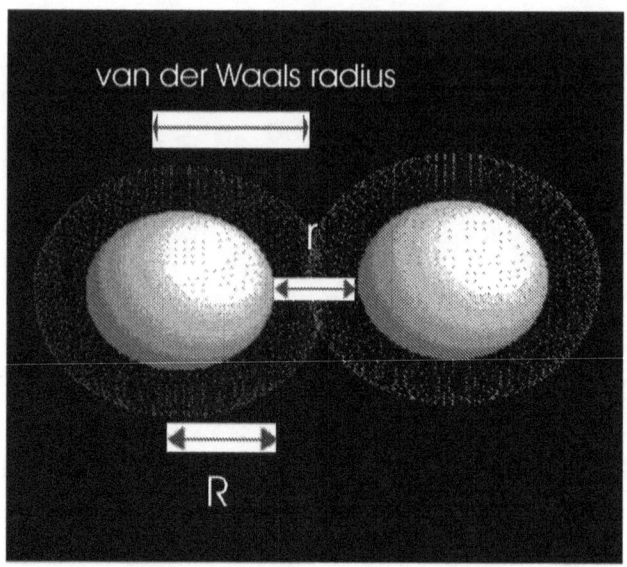

Figure 4. Definition of molecular separation (r), molecular radius (R), and van der Waals radius.

van der Waals radius $= t_{van} = R + r / 2, A^0$ (6)

Volume of a single molecule $= v_{cal} = 4/3 \, \pi \, t_{van}^3 \, (A^{03}) = 4.189 \times t_{van}^3 \, (A^{03})$ (7)

Number of molecules in one $cm^3 = n = 1 \times 10^{24} \, (A^{03}) / v_{cal} \, (A^{03})$ (8)

Mass of n molecules $= M = n \times$ mass of one C_{60} molecule (9)

$= n \times 1196.728 \times 10 \times 10^{-24} \, (g)$ (10)

Density = mass / volume = M / V $= M / (1 \, cm^3)$

$= n \times 1196.728 \times 10 \times 10^{-24} \, (g / cm^3)$ (11)

$= 1196.728 / v_{cal} \, (g / cm^3)$ (12)

Equation (12) is plotted against intermolecular separation (r) in Figure 5 for various R shown on the graph.

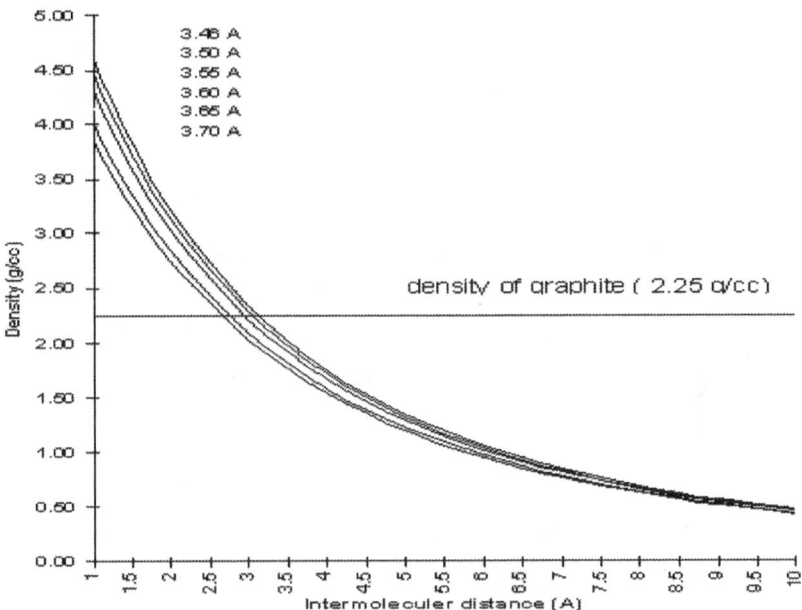

Figure 5. Plot of density (g/cc) vs. molecular separation according to equation (12) for various molecular radii shown on the graph.

It is seen that the density decreases exponentially as r increase. The effect of R on density is more pronounced when r is small. However, R becomes less influential as r becomes larger. At this point, we may estimate the upper limits to R and r using the density of graphite as the upper limit. If we set C- C bond length as 1.41 A^0 (average of C_5-C_6 and C_6-C_6 reported in part I), then we calculate R $\approx$3.0 A^0. Using this value, we predict that r $\approx$ 3.0 A^0. The latter value is in agreement with our prediction for trimer model.

The density data may be further used to predict the magnitude of r, or at least the upper limit. As the calculated density of solid C_{60} is lower than the graphite density (or the compressibility is higher for graphite), we assume that the graphite density sets the uper limit to r. Thus, the upper limit for r was computed to be about 3.0 A^0 combining the density of the graphite and the above equations(6-12), which is in agreement with our calculated value for timer model. This might perhaps suggest that the calculated density of 1.18 g/cm^3 for bulk C_{60} might be an upper limit rather than its actual value.

Acknowledgment: The graphics in Figures 1, 2 and 4 are done with the aid of RasMol. I sincerely thank Roger Sayle, Bimolecular Structure Department, Glaxo Wellcome Research Development, Stevenage, Hertforshire, UK, for providing this graphics tool.

Reference

1.   For examples see, (a) Hammond G.S. and Kluck V.J. Eds., *Fullerenes*, ACS Symposium 481; American Chemical Society, Washington, D.C., 1992 ; (b) Special Issue On *Buckminsterfullerence, Acc. Chem. Res.* 1992, 25(3); (c) Billups W.E. and Ciufolini, M.A.,Eds.,*Buckminsterfulleres*, VCH Publishers, Inc., N.Y.1993, and the references cited therein.

2.   Gorum S.M., Greaney M.A., Day V.W., Day C.S., Upton R.M. and Briant, C.E., in *Fullerenes*, Hammond G.S. and Kluck V.J. Eds., American Chemical Society, Washington, D.C., 1992, chapter 3.

3.   Hawkins J.M., Meyer A. Lewis, T.A. and Loren, S., in *Fullerenes*, Hammond G.S. and Kluck, V.J. Eds., American Chemical Society, Washington, D.C., 1992 , chapter 6.

4.   Fagan P.J., Calabrese J.C. and Malone B., *Acc. Chem. Res.* 1992, 25, 134.25.; Fagan P.J., Calabrese J.C. and Malone B., in *Fullerenes*, Hammond G.S. and Kluck, V.J. Eds., American Chemical Society, Washington, D.C., 1992 , chapter 11.

5.   Yannoni C.S., Bernier P.P., Bethune D.S., Meijer G. and Salem R.J., *J.Am.Chem.Soc.* 1991, 113, 3190; Johnson R.D., Bethune D.S. and Yannoni C.S., *Acc. Chem. Res.* 1992, 25, 169; Cox D.M., Sherwood R.D., Tindall, P, Creegan K.M., Anderson, W. and

Martell D., in *Fullerenes*, Hammond G.S. and Kluck, V.J. Eds., American Chemical Society, Washington,DC, 1992, chapter 8.; Cox D.M., Behal K., Creegan, K., Disko M., Hsu C,S., Kollin E., Miler J., Robbins J., Robbins W., Sherwood R.D, Tindall P., Fisher D. and Meitzer G., *Proc. Mater. Res. Soc.* Boston, 1991, 206, 651.

7. Weltner W., Jr. and Van Zee R.J., *Chem. Rev.* 1989, 89, 1713.

8. Scuseria G.E., *Chem.Phys. Lett.* 1991, 176, 423.

9. Kumbar S. and Kumbar M., part I of this series.

10. Fischer J.E., Heinley P.A. and Smith III A.B., *Acc. Chem. Res.* 1992, 25, 112.

11. Fischer J.E., Heiney P.A., Luzzi D.E. and Cox D.E., in *Fullerenes*, Hammond G.S. and Kluck, V.J., Eds., American Chemical Society, Washington, D.C., 1992, chapter 4.

12. Fleming R.M., Hessen B., Siegrist T., Kortan A.R., Marsh P., Tycko R., Dabbagh G. and Hadden R.C., in *Fullerenes*, Hammond G.S.and Kluck V.J.Eds., American Chemical Society, Washington, DC, 1992, chapter 2.

13. Hawkins J.M.,*Acc. Chem. Res.*, 1992, 25, 150.

14. Wilson R.J., Meijer, G., Bethune D.J., Johnson R.D., Chamblis D.D., De Vries M.S., Hunziker H.E. and Wendt H.R., *Nature(London)*, 1990, 348, 621.

15. Wragg J.L., Chamberlin J.E., White H.W., Kratschmer W. and Huffman D.R. *Nature(London)*, 1990, 348, 623.

16. Holczer K., Klein O., Huang S.M., Kaner R.B., Fu K.J., Whetten R.L. and Diederich F., *Science* (Washington D.C.) 1991, 252, 1154.

17. Fleming R.M., Siegrist T., March P.M., Hessen B., Kortan A.R., Murphy D.W., Hadden R.C., Tycko R., Dabbagh G., Mujsce A.M., Kaplan M.L. and Zahurak S.M., *In Cluster-Assembled Materials*; Averback R.S., Bernole J. and Nelson D.L.,Eds., *Material Research Society Symposium Proceedings* 206, Material Research Society, Pittsburgh, PA, 1991, p 691.

18.   Tycko R., Hadden R.C.,Dabbagh G., Glarum S.H., Douglas D.C. and Mujsce A.M. *J.Phys.Chem* 1991, 95,518.

19.   Kumbar M., *J. Theor. Biol,* 1975, 97, 7411.

20.   Kumbar M. and Maddaiah V.T., *Biochem.Biophys.Acta.*, 1977, 497, 707.

21.   Murrel J.N., Kettle S.F.A. and Tedder J.M., *The Chemical Bond*, John Wiley and Sons, N.Y., 1978, p 288.

22.   Kauzman W.,*Quantum Chemistry,* Academic Press Inc., New York 1957.

23.   Margenau H. and Kestner N.R., *Theory of Intermolecular Forces*, Pergamon Press, N.Y. 1969, p 27.

24.   Press W.H., Flannery B.P., Teukolsky S.A. and Vetterling, W.J.. *Numerical Recipes, The Art of Scientific Computing*, Cambridge University press, N.Y. 1986, chapter 10.

25.   Selby S.M. and Girling, B.; Eds., *Standard Mathematical Tables*, The Chemical Rubber Co., Cleveland, Ohio, 1965, p 491.

Appendix D
Solution : Molecular Point of View and Molecular Weighing Balance

Mahadev Kumbar (1992)

What is a solution? A solution is the homogeneous mixture of a solute and a solvent. Of course, this is the definition of a simple solution consisting of one solute and one solvent or it may be simply called as a '**two-component system**. Within the boundary of such simple solution, one might define various categories of solutions. Among them, dilute and concentrated solutions are important to us because some of these are considered to be the stock solutions, usually kept in the general chemistry laboratory. The terms 'dilute' and 'concentrated', as we use them so freely, can be attributed to their relative strength; we need to compare one solution with another solution in order to label the solution in question as either a dilute or a concentrated solution (Generally, the concentrated solution refers to the solution that contains the maximum amount of solute under normal conditions). For example, we say that 6M (six molar) HCL (hydrochloric acid) is a dilute solution or 12M (twelve molar) HCL is a concentrated solution. In a real sense, 6M solution is a dilute solution compared to 12M solution or 12M solution is a concentrated solution compared to 6M solution. The degree of diluteness (or concentrateness) is embedded in how much 6M (or 12 M) solution is weaker (stronger) than a particular concentration. For example, if we talk about 7M or 8M solution, most probably we would say that it is more concentrated than 6M solution but we would not say that it is a concentrated solution. Therefore, one can see that certain amount of ambiguity is always brought about when using the terminology such as 'dilute' or 'concentrated.' We tell students that HCL is kept in the laboratory in two concentrations , namely, the dilute form (6M) and the concentrated form (12M). If students ask the question, such as, *what do you mean by that 12M is concentrated and 6M is dilute ?* Well, how should we explain it ? Of course, we can always explain that 12M solution contains more solute particles (or less number of water molecules) than 6M solution (or more number of water molecules) in a fixed volume of solution. At this point, students might pop-up another question, like, *what do you mean by that the concentrated solution contains more solute particles than dilute solution , exactly how many more, and how can we see or verify this?* Now, I think we are really stuck. The only way out of this quandary is to explain the difference between 12M and 6M solutions in terms of number of solute species and/or number of solvent molecules present in both solutions. That is to say that we have to convert macroscopic explanation (solution in terms of molarity) into microscopic explanation (solution in terms of ions, molecules, etc.).

Chemistry is the science of molecules and atoms and is the query of chemical laws and forces. When chemical reactions take place, they do so on molecular or atomic level - molecules first break down into atoms, atoms reshuffle, and finally molecules reform by the chemical union of atoms - this is the pathway for most inorganic chemical reactions. But, when we wish to carry out the chemical reactions in the laboratory, especially, in the general chemistry laboratory, we weigh certain amount of substance A and certain amount of substance B in grams (usually) or in any other suitable unit. We know for the fact that the mass as a whole entity in its entirety does not react, but its individual constituents like molecules and atoms, existing within the domain of the weighed mass react. In explaining the mechanics of the chemical reactions to students, we write chemical reactions in the form of chemical equations, and use moles or molecules to explain the process of the chemical reaction. When it is time for doing the experiment, we tell students to weigh the substance in grams not in molecules or atoms. **This is because a balance for weighing molecules or atoms is not currently available.** Of course, one can always convert the mass into molecules or atoms, which might be tedious on students' part . **Instead, if the balance shows the amount of substance as well as its equivalent number of molecules or atoms, then I feel that it would not only be helpful to students but also it would give them the better understanding of the chemical reactions.** Chemical reaction may be thought of as a kind of war between participating entities. For examples, if we say that there is a war between two battalions, it really does not give us the clear picture of the war. But, if we say that there is war between two battalions consisting of 10,000 soldiers in each, it gives us not only the clear picture of the war but also its enormity. At present, students can see only the mass (like the battalion in the above example) when they weigh the substance on a balance. In addition, if they could also see corresponding number of molecules or atoms(like soldiers in the above example), then I am optimistic that they would really appreciate the chemical reaction due to the fact that they could visualize the number of individual entities participating in the chemical reaction.

Molecular weighing balance may also bring some uniformity in weighing objects. At present, it appears that there exists some inconsistency in our weighing of substances of different physical nature ; solids are weighed on a balance, liquids are either weighed on a balance or their volumes are measured, and gases are not weighed on a balance but their pressures (volumes) are measured. Regardless of their physical state of matter, molecules or atoms always participate in chemical reactions. **Therefore, such inconsistency can be eliminated by adopting a common procedure by designing the balance that can display the number of molecules or atoms.** After all, our ultimate aim is to have students to get more interested in conceptualizing or visualizing the chemical reactions in terms of its participating molecules or atoms - a path that may lead towards the better understanding of chemistry.

Chemistry, as it has progressed in the past from one century to the next century, has provided us with deeper and deeper understanding of the nature on microscopic level. Human eyes, on the other hand, without an external assistance (like instruments), can see the nature only on macroscopic level - phenomenon characteristic to human eyes but very frustrating to chemists. As we insist on seeing the microscopic nature with our bare eyes without any external help, we must learn to do things in a different way. Conceptualizing chemistry on a microscopic level may not be difficult for a trained chemist , but it is difficult for students. Can we make students see or even conceptualize molecules and atoms in their test tubes? May be or may be not. But, at least, we can stretch their imagination beyond the macroscopic level. A simple example will suffice to illustrate this. Let us ask students a very simple question: *if they drink a small glass of water (say about 250 ml), how many water molecules are they drinking?* To answer this question, let us calculate number of water molecules in the given glass of water.

$$\text{No. of water molecules} = (250 \text{ ml}) \times (0.99079 \text{ g/ml}) \times (1/18.0 \text{ g/mol})$$
$$\times \ 6.023 \times 10^{23} \ (\text{molecules/mol})$$
$$= 8.3 \times 10^{24} \text{ water molecules.}$$

We can now tell students that when they drink a glass of water, in fact, they swallow 8,300,000,000,000,000,000,000,000 water molecules. Suppose we wish to arrange these water molecules next to each other in a row, the row will extend up to about 1.2×10^{12} miles by assuming that the length of the water molecule about 2.3 A^0 [$(8.3 \times 10^{24}$ water molecule) $\times (2.3$ A^0/water molecule) $\times (10^{-8}$ cm/A^0) $\times$ (1 in/ 2.54 cm) $\times$ (1ft / 12 in) $\times$ (1mi / 5280 ft) $= 1.2 \times 10^{12}$ mi], which is bout 5 million times the distance between the moon and the earth (the distance between the moon and the earth is about 240,000 mi). Our inability to count such large number of water molecules dictates us to say that 'a glass of water.'

From this examples, it is intuitive that we can see things much more clearly if we analyze the solution in terms of atoms and molecules. In the following, I have attempted to describe the procedure and methodology that students can use to gain some insight into a solution or possibly a manufacturer to design molecular *weighing balance.*

Calculation of Number of Ions and Water Molecules
(a) **Ions.** Let n_i be the number of moles of the ith ion in the given solution. The total number of ith ions, N_i, is given by,

$$N_i = n_i \times N_A.$$ \hfill (1)

where N_A is the Avogadro's number (6.023×10^{23}). The total number of ions in solution , N_T , is the sum all the ions, which is expressed as,

$$N_T = \Sigma \ N_i = \Sigma \ n_i \times N_A .$$ \hfill (2)

(b) **Water molecules.** Number of water molecules present in a given solution can be estimated by two method, depending upon the information available:

(i) *Density Method.* If the solution density is available, the number of water molecules can be obtained by the following equation.

$$n_{water} = [\ \{ \ (V_{soln} \times d_{soln}) - (M_{soln} \times V_{soln} \times m_{solute}) \ \} \ / \ m_{water} \ ] \times N_A$$ \hfill (3)

where V_{soln} = volume of the solution in ml or L

d_{soln} = density of the solution in g/ml

M_{soln} = molarity of the solution in mol/l

m_{solute} = molar mass of the solute in g/mol

m_{water} = molar mass of water in g/mol (= 18.00 g/mol)

The first term in equation (3) represents the mass of the solution, the second term the mass of the solute, and the term in the square bracket the number of moles of water.

(ii) *Ionic Radius Method.* If the solution density is not available, the number of water molecules can be estimated using the ionic radius. To keep the calculations simple we assume that the ions are hard spheres. Let r_i be the radius of the ith ion in nm. Its volume (nm^3), $v_{i,}$, is determined by

$$v_i = 4/3 \ \pi \ r_i^3 \tag{4}$$

where π is the constant ($= 3.14159$) . The above expression yields the volume of a single ion. However, the volume of m_i moles of ith ion, v_i^m, is calculated by the following equation.

$$v_i^m = v_i \ x \ m_i \ x \ N_A \ nm^3 \tag{5}$$

The total volume of all the ions in the solution, V_{ion} , is the sum of volumes of all the ions, that is,

$$V_{ion} = \Sigma v_i^m \ nm^3 \tag{6}$$

The volume of the water, V_{water} , is now obtained by subtracting V_{ion} from the volume of the solution, V_{soln}, that is,

$$V_{water} = V_{soln} - V_{ion} \tag{7}$$

The problem can be further simplified by assuming that the volume of the solution is equal to one liter, that is , we set $V_{soln} = 1 \ L$ ($= 1 \ x \ 10^{24}$ nm^3). Under this condition the equation (7) becomes

$$V_{water} = 1 \times 10^{24} \text{ nm}^3 - V_{ion} \tag{8}$$

It should be noted that equation(8) gives the volume in nm^3. This is necessary because most standard general chemistry text books report the ionic radius in nm unit. The total number of water molecules, n_{water}, in V_{soln} is computed with the help of the following equation.

$$n_{water} = V_{water} \times t_{water} \tag{9}$$

where t_{water} is the number of water molecules in one nm^3 volume, which can be evaluated knowing the density of water d_{water} (=0.997099 g /ml) and the molar mass of the water through the following equation.

$$\begin{aligned} t_{water} &= (d_{water} / m_{water}) \times N_A \times 10^{-21} / \text{nm}^3 \\ &= 33.464 \text{ molecules/nm}^3 \end{aligned} \tag{10}$$

This number, of course, represents the number of water molecules present in one nm^3 volume based on the above selected density for water. The above described procedures [equations(1) - (10)] to compute number of ions and number of water molecules is illustrated below using 6M HCL as an example.

Example. Let us consider 6M HCL solution and assume 1L solution. The HCL is considered to be a strong electrolyte and due to that its ionization reaction is given by the following chemical equation.

		HCL	$\rightarrow$	H$^+$	+	CL$^-$
n_i	=	6M		6M		6M

Number of moles of H^+ ions $= n_H = 6$ mol
Number of moles of CL^- ions $= n_{CL} = 6$ mol
Number of H^+ ions $= N_H = n_H \times N_A = 6$ mol x 6.023×10^{23} H^+ ions/mol $= 3.61 \times 10^{24}$ H^+ ions
Number of CL^- ions $= N_{CL} = n_{CL} \times N_A = 6$ mol x 6.023×10^{23} CL^- ions/mol$= 3.61 \times 10^{24}$ CL^- ions
Total number of ions $= N_T = N_H + N_{CL} = 3.61 \times 10^{24} + 3.61 \times 10^{24} = 7.22 \times 10^{24}$ ions

Calculation of H_2O molecules
(i) Density Method.
Volume of solution $= V_{soln} = 1L = 1 \times 10^3$ ml
Density of 6 M HCL $= d_{soln} = 1.10$ g/ml
Molarity of the solution $= M_{soln} = 6M$
Molar mass of the solute $= m_{solute} = 36$ g/mol
Molar mass of water $= m_{water} = 18.0$ g/mol

Substituting these values into equation(3), we obtain the number of water molecules in 6M HCL solution:

n_{water} $= [\{(1 \times 10^3 \, (ml) \times 1.10 \, (g/ml) - (6 \, (mol/L) \times 1L \times 36 \, (g/mol)) \}/ 18.0 \, (g/mol)]$
$\times 6.023 \times 10^{23}$ (molecules/mol)
$= 2.95 \times 10^{25}$ H_2O molecules   (11)

(ii) Ionic Radius Method.
The radius of $H^+ = 0.208$ nm and $CL^- = 0.181$ nm. Substituting these values into equation (4) yields the volumes of these ions, which are

$v_H = 4/3 \, \pi \, (0.208)^3 = 3.77 \times 10^{-2}$ nm^3
$v_{Cl} = 4/3 \, \pi \, (0.181)^3 = 2.48 \times 10^{-2}$ nm^3

Volumes of 6 molar ions, according to the equation(5), are

Volume of 6 molar H^+ ions $= v_H^{6m}$ $= 6 \, (mol) \times (6.023 \times 10^{23}/ mol) (3.77 \times 10^{-2} \, nm^3)$
$= 1.362 \times 10^{23}$ nm^3
Volume of 6 molar CL^- ions $= v_{Cl}^{6m}$ $= 6 \, (mol) \times (6.023 \times 10^{23}/ mol) (2.48 \times 10^{-2} \, nm^3)$
$= 8.962 \times 10^{22}$ nm^3

Total volume of ions is calculated using equation (6) as
$$V_{ions} = v_H^{6m} + v_{CL}^{6m} = 2.26 \times 10^{23} \text{ nm}^3$$
The volume of water according to equation (8) is
$$V_{water} = 1 \times 10^{24} \text{ nm}^3 - 2.26 \times 10^{23} \text{ nm}^3$$
$$= 7.742 \times 10^{23} \text{ nm}^3$$
$$= 774.2 \text{ ml}$$

Thus 6M HCL contains about 774 ml water. To obtain the number of water molecules (equation (9)), 7.742×10^{23} nm^3 is multiplied by 33.464 molecules / nm^3 (equation (10)). Hence, we have

$$n_{water} = 7.742 \times 10^{23} \text{ nm}^3 \times 33.464 \text{ molecules / nm}^3$$
$$= 2.59 \times 10^{25} \text{ H}_2\text{O molecules} \tag{12}$$

It is apparent that the density method and the ionic radius method do not yield exactly the same value for number of water molecules (compare Equation(11) and (12)). This discrepancy might be attributed to the spherical approximation made in regard to ions in the ionic radius method, which seems to overestimate the ionic volumes. However, one should be aware of the fact that the density method always yields more accurate results than the ionic radius method provided accurate densities are available.

Results and Discussion

The above presented procedure has been applied to analyze some common stock solutions kept in the laboratory. The procedure is general and can be extended to any solution regardless of its nature, whether it is weak or strong, as long as the number of moles of ions in solutions is known. Table I summarizes number of parameters for various stock solutions in dilute and concentrated forms. It is not possible to compare one stock solution (for e.g. HCL) with another stock solution (for e.g. HNO_3) either for dilute or concentrated solution due to different molarities involved.

Table I Summary of number of species for various common stock solutions.

	Solution Conc.	Density	H⁺/OH⁻ ions	Anion/cation	Undissociated. molecule	H₂O molecules
	(M)	(g/ml)	x 10^{-26}	x 10^{-24}	x 10^{-24}	x 10^{-25}
CH₃COOH	6	1.04	6.02 x 10^{-3}	6.02 x 10^{-3}	3.61	2.27
	17	1.18	1.05 x 10^{-2}	1.05 x 10^{-2}	10.23	0.10
HCL	6	1.10	3.61	3.61	0	2.95
	12	1.18	7.22	7.22	0	2.50
HNO₃	6	1.19	3.61	3.61	0	2.43
	16	1.42	9.63	9.63	0	0.63
H₂SO₄	3	1.18	3.62	1.81	0	2.96
	18	1.84	21.68	10.84	0	2.54
NH₄OH	6	0.96	6.02x10^{-3}	6.02x10^{-3}	3.61	2.5
	15	0.90	9.88x10^{-3}	9.88x10^{-3}	9.02	1.25
NaOH	6	1.22	3.61	3.61	0	3.28
	14.3	1.43	8.61	8.61	0	2.87

*includes ions, undissociated molecules, and water molecules

Table I. Continued.

Solution	Total species* x 10^{-23}	%H⁺/OH⁻ ions	%Anion/cation	% Undissoc. molecule	% H₂O
CH₃COOH	2.63	2.28x10^{-2}	2.28x10^{-2}	13.69	86.26
	1.12	9.36x10^{-2}	9.36x10^{-2}	90.89	8.91
HCL	3.68	9.82	9.82	0	80.35
	3.16	18.31	18.31	0	63.37
HNO₃	3.16	11.45	11.45	0	77.10
	2.56	37.70	37.70	0	24.59
H₂SO₄	3.50	10.31	5.16	0	84.53
	3.51	61.84	30.92	0	7.25
NH₄OH	2.87	2.09x10^{-2}	2.09x10^{-2}	12.57	87.38
	2.16	4.58x10^{-2}	4.58x10^{-2}	41.81	58.09
NaOH	3.99	9.04	9.04	0	81.93
	4.59	18.76	18.76	0	62.48

However, it is possible to compare the same stock solution in dilute and concentrated forms. As one might expect, number of H^+ ions(acidic solutions) or number of OH^- ions(basic solutions) is greater for concentrated solutions compared to dilute solutions. It is not clear how the % ions varies in general with molarity by just two calculations. To have some idea about this behavior, at least for a simple acid like HCL, % H^+ ions and % total ions are computed for nine molar concentrations (Table II) and are plotted in Figure 1.

Table II. Percent of H^+ ions and percent of total ions for HCL at various molarities.

Molarity(M)	% H^+ ions	% Total ions
0.137	0.246	0.492
1.402	2.464	4.930
2.872	4.940	9.860
4.472	7.396	14.793
6.022	9.861	19.722
8.047	12.818	25.635
9.454	14.790	29.580
11.639	17.746	35.493
13.139	19.718	39.435

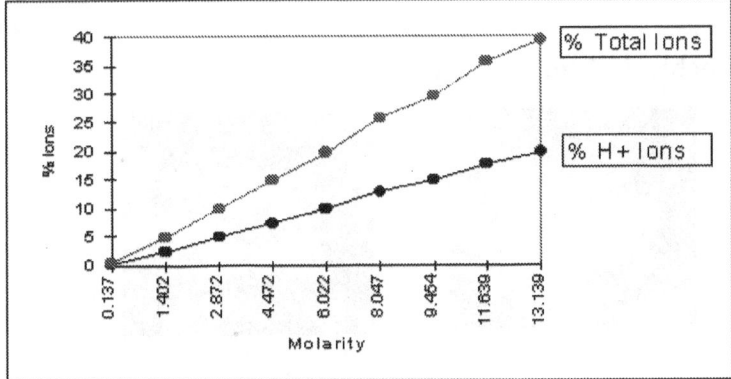

Figure 1. Plot of % H^+ ions and % total ions($H^+$and OH^-) vs. molarity for HCL solution.

It is evident that both % H$^+$ ions and % total ions increase with the molarity in somewhat non-linear fashion, and may be fitted to the following equations:

% H ions   =   Molarity / [(0.0091 x Molarity) + 0.556]
% Total ions =   Molarity / [(0.0045 x Molarity) + 0.278]   (13)

It is more intuitive to transform the information in Table I into a visual kind of display (Figure 2).

(a) Acetic Acid (HC$_2$H$_3$O$_2$(aq))

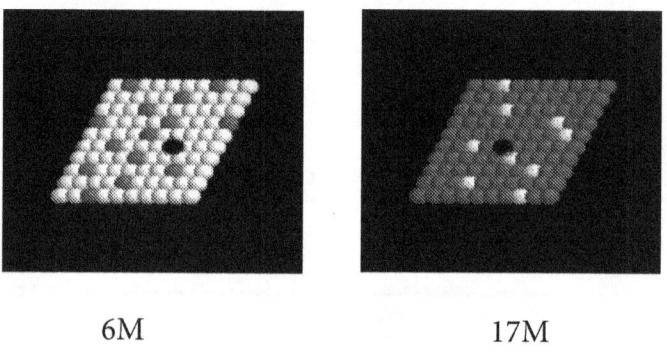

6M   17M

(b) Hydrochloric Acid (HCL(aq))

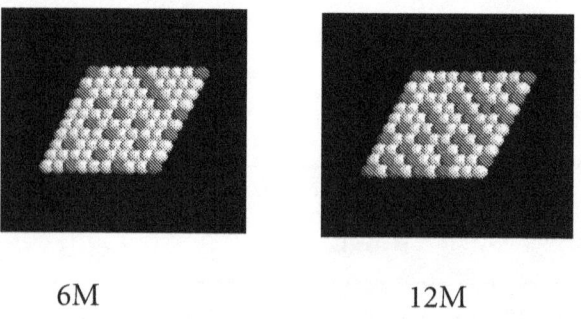

6M   12M

(c) Nitric Acid(HNO₃ (aq))

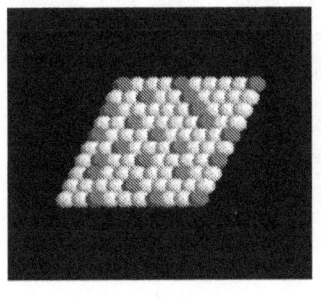 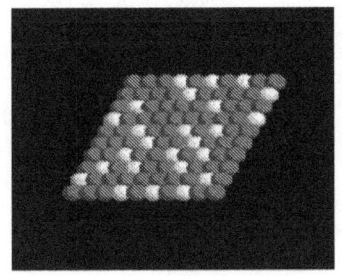

6M   16M

(d) Sulfuric Acid (H₂SO₄ (aq))

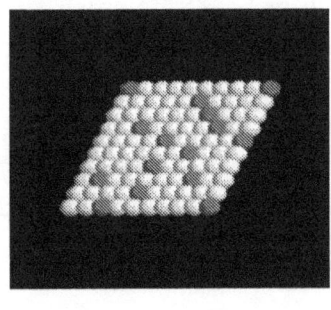 

3M   18M

(e) Ammonium Hydroxide (NH₄OH (aq))

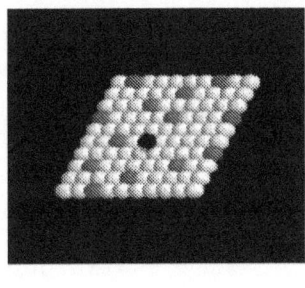 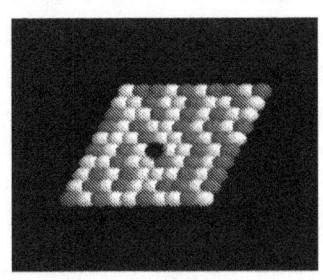

6M   15M

(f) Sodium Hydroxide (NaOH (aq))

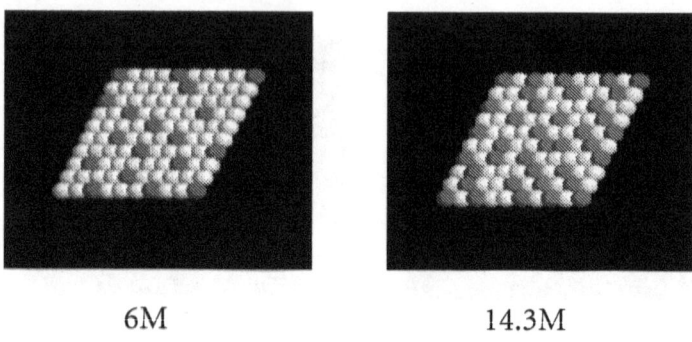

6M   14.3M

Figure 2. Schematic representation of %ion (gray ball), % undissociated molecules (gray balls in case of acetic acid and ammonium hydroxide), and % water molecules (white balls) for stock solutions. There are total of 100 circles. Even though the % ions in acetic acid and ammonium hydroxide is much less than 1%, one black ball is used to represent symbolically the presence of ions: (a) Acetic acid (6M and 17M), (b) Hydrochloric acid (6M and 12M), (c) Nitric acid (6M and 16M), (d) Sulfuric acid (3M and 18M), (e) Ammonium hydroxide (6M and 15M), and (f) Sodium hydroxide (6M and 14.3M).

The total number of species is equal to 100, as it is easier to calculate percentages. I feel that these diagrams clearly display solutions on molecular level giving a better visual effect of dilute and concentrated solutions. It is my experience that such representation is more informative than explaining students using molar concentrations alone. I am quite optimistic that students will grasp the concept of solution much quicker and easier if they view the solution molecular or atomic point of view instead of molar point of view.

Acknowledgment: The above graphics (Fig.2) are done with the aid of RasMol. I sincerely thank Roger Sayle, Bimolecular Structure Department, Glaxo Wellcome Research Development, Stevenage, Hertforshire, UK, for providing this graphics tool.

Appendix E
A Brief History of Chemistry

Introduction

I   Pre-Alchemy (2 million BC – 600 BC)

II   Alchemy (600 BC- Mid 17th Century)
     IIa Material (Exoteric) Alchemy
     IIb Spiritual (Esoteric) Alchemy
     IIc Iatrochemy or Medicinal Alchemy

III Post-Alchemy (Mid 17th Century – Today)
     IIIa Birth of Chemistry
                    III.a.a   Early Knowledge
                    III.a.b   The Oxford Trios
                    III.a.c   The Theory of Phlogiston

     IIIb Foundation of Modern Chemistry
                    IIIb.a   Electrochemical or Dualistic Theory
                    IIIb.b   The Radical Theory
                    IIIb.c   Theory of Substitution and Other
                                Theories
                    IIIb.d   The Theory of Valence

IIIc Renaissance in Chemistry

IIIc.a	The Rise of Organic Chemistry
IIIc.b	The Rise of Physical Chemistry
IIIc.c	The Rise of Inorganic Chemistry
IIIc.d	The Rise of Biochemistry
IIIc.e	The Structure of an Atom and Chemical Bonding

References

Introduction

Looking at the kind of science the chemistry is, it appears that the history of which has no beginning as chemical phenomena and chemical reactions have been taking place in nature even before the Big Bang – a series of explosives and events resulting in a hot, dense, and expanding matter and energy, which is known as the Big Bang Theory. According to this theory, these explosive events suppose to have taken place about 10 or 15 billion years ago. Since the matter was involved in the explosion, where it came from? How was it formed before the Big Bang? And what was the nature of this matter before the Big Bang? These are some of the questions yet to be answered. These series of explosives during the Big Bang may be conceptualized as nothing more than highly explosive chemical reactions (combustion reactions) or nuclear explosive reactions that may be explained by chemical principles if the reacting matter is known precisely.

As the universe cooled as it expanded-according to Big Bang Theory-protons (hydrogen nuclei (H)) are formed within the first second. In the next 1,000 seconds or so, protons and neutrons are combined to form deuterium (D) nucleus, the isotope of hydrogen. During this brief period of time, some other light nuclei like helium (He), lithium (Li), beryllium (Be), and boron (B) are also formed. Come to think of it, the origin of nuclear chemistry can be attributed to Big Bang, and therefore, the nuclear chemistry may be considered as the first science of chemistry. Since that time, chemical reactions of one or the other kind have been taking place continuously in our universe, leading towards the current status of our universe.

When we talk about the history of chemistry, we usually talk about its association with human beings in one or the other form. In that respect, the history of chemistry is very fascinating. Its flight to the modern chemistry is covered with speculations, quirks, witchcrafts, and paradoxes. The science of chemistry dates back to Neanderthals (lived about 24,000 years ago) who used fire to keep themselves warm in wintertime and also keep wild animals from entering the caves. Later on, wines were produced, potteries were made, and cave walls were painted. Eventually, chemical reactions were invented to make metals, glass, and other kinds of materials needed to sustain the life necessities. But these were made by trial and error method and also by word-to-mouth. No attempt was made to understand the chemical behavior or organize the thoughts as a systematic science until about twenty-five hundred years ago, when the Greeks began to organize thoughts into a systematic science. However, their concept of matter slowly led many philosophers (so called thinkers) away from the science of chemistry, and finally, the Greek science lost its influence about four hundred years ago making a room for the start of birth of early chemistry.

If we scrutinize the history of chemistry, it appears that alchemy played a crucial role in attempting to recognize the science of chemistry. Ideas of alchemy were good, but their implementations, procedures and methodologies lased with spiritual dogma in doing so, were not correct. We can say that the earlier chemistry was born as resentment towards alchemy. Yet, it is ironic to note the connection between them - many procedures, chemicals, and apparatus were the same. Even the names and terminology were often similar. Not only that but also many earlier chemists were also alchemists. Keeping this prospective in mind, I have divided the history of chemistry into three sections:

I   Pre-Alchemy (2 million BC – 600 BC)
II   Alchemy (600 BC – Mid 17th Century)
III   Post-Alchemy (Mid 17th Century – Today),

which are discussed below.

I   Pre-Alchemy (2 million BC – 600 BC)

It took millions of years for human evolution to evolve into human (*homo sapiens*), the course of which was charted by many forces like climatic conditions, green pasture, availability of food, and other natural resources, and above all, the breadth of knowledge about the nature. The evolution of chemistry also took place along the side of human evolution, often prompted by the need for one or the other kind. Hence, human evolution and the evolution of chemistry are complimentary to each other.

In the beginning of the Old Stone Age or *Paleolithic* period (2 million – 13,000 BC, corresponded roughly to the Ice Age that began millions of years ago and ended about 8000 BC), humanoid creatures were the first one to use stone tools for food gathering through hunting, as it was their main occupation. Over time, a variety of tools from stone were made for specific purposes. By more than 100,000 years ago, Neanderthals had developed several types of tools using stone as well bone or horn. At the end of *Paleolithic* period, modern human (*homo sapiens sapiens*) were able to make such specialized tools as slender small objects, like needles and harpoons.

Paintings and engraving on the walls of the caves, dating back to 25,000 to 10,000 years ago, have been found in Altamira in Spain and Lascaux in France. Mineral pigments mixed with animal fat were used in multicolor animal figures. The animal fat appears to have been used as a binding agent. Artists most likely have used the pigments available locally at that time. They used four colors: the black pigment made from charcoal or from manganese consisted of iron oxide found in ores and clays; red and yellow, made from ocher, a kind of impure iron ore; and white, made from pipe clay. Thus, the chemistry of paints and painting was born. The idea of colors might have been derived from observing the nature. And thus the extraction of colors from nature, eventually transformed into the chemistry of natural products during the post-alchemy era. Since inner chambers of caves were painted, most likely oil lamps or torches were used as a source of light indicating the start of chemical reactions by burning some kind of natural substances. During this era also sorcery, magic, and some form of religious activities were emerged, which were later to become part of chemistry during Alchemy period.

During the Middle Stone Age or *Mesolithic* period (13,000 BC – 8,000 BC), weather became warmer due to retraction of last glaciers of the Ice Age, as a result, forests were replaced by the open plains of the Old Stone Age. *Paleolithic* tools were further improved and adopted to new environmental conditions. The hunting, which was the main occupation in the Old Stone Age, probably became more difficult and hence alternate ways of gathering foods were invented, like, cultivating plants and domesticating animals.

Agriculture slowly began to emerge during the *Neolithic* period (8,000 BC -) in the Middle East and in Mesoamerica. Stone and bone tools became more varied and sharpened. By 6,000 BC pottery appeared in the ancient Middle East, and copper was first in some regions. People began to settle in villages, houses were built using stones, straws, or even primitive bricks. They painted and/or scratched designs on the pottery before baking it in fire, made baskets, spun wools, cotton, and flax, and wove cloth.

Religion or the blind belief became more prominent due to agricultural life. They knew the importance of rain and sun for yielding the crops. The earth was worshipped as mother Goddess for better crops. Other Gods were invented to represent the rain and the sun. In scarcity of rain and sun, rituals were performed and sacrifices were offered to please rain-God and sun-God. These beliefs led to rituals and eventually creating a special class of people known as "priests."

As far as we know, recorded history began about 3000 BC. By this time, some villages in Mesopotamia (present-day Iraq) and Egypt had been transformed into cities. Writing was invented. Smelting and casting of copper, silver, and gold into ornaments, and making tools and weapons began. Shortly after 3000 BC, the bronze -an alloy - was produced by mixing copper with tin, starting a new age called "Bronze Age." This is an indication that artisans knew how to raise the temperature of a furnace or clay pit high enough to melt these metals (melting point of copper is 1083^0 C and that of tin is 232^0 C) and to alter the properties of pure metals. This is very important step because it started a whole new series of metallurgical chemistry. The Iron Age followed the Bronze Age when people first smelted and forged iron at about 1100 BC in Asia Minor (now Turkey).

Most of the artisans were employed by kings and temples to make luxury goods for nobles and priests. Priests in the temple had ample time and luxury to speculate on the cause and the origin of changes taking place in the world surrounding them. Their hypothesis often involved magic or some sort of sorcery, but they also formulated ideas about astronomical and cosmological events, and mathematics concepts.

The idea that nature is created with few fundamental elements appears to have been originated in India or China or both places simultaneously as described in Hindu and Chinese scriptures. About 1400 BC or earlier in India, *Vedas* (meaning "knowledge," are considered to be the lighthouse of eternal wisdom leading towards salvation) written in Sanskrit language (one of the oldest language dates back to 17th century BC at least) mentions five elements, earth, water, air, ether or space, and fire. In addition, *Rig-Veda* mentions gold, silver, copper, lead, bronze and possibly iron and tin. However, The Upanishads (900- 500 BC) definitely mentions iron. In addition, *Rig-Veda* also refers to *Soma sacrifice*, the main ritual activity. The *Soma* was a hallucinogenic beverage prepared from some kind of a plant like mushroom or fungus, but cannot be said with certainty. In any event, it indicates that some kind of chemical procedure was known to Indians at that time.

In ancient China about 1200 BC "book of changes" (*I Ching*) ascribed to Wen Wong introduces duality principle *yin* and *yan*, perhaps associated with dark and light or female and male respectively. Whether they were considered to be cosmic forces or material elements is very uncertain.

II   Alchemy (600 BC- Mid 17th Century)

When we think of alchemy, we think of fake pursuit of transmutation of base metals such as lead into gold. This is traditionally known as alchemy or material (exoteric) alchemy that dominated Mesopotamia, Egypt, Greece, Asia Minor, and rest of the Europe. There is, however, another kind of alchemy that concerned with the spiritual world, which is known as spiritual (esoteric) alchemy that involved looking for "elixir of life" originated in India or China or both. Although, no success was achieved in either converting base metals into gold or come up with elixir of life, the amount of work put into did not go waste, rather many new substances like alcohol, mineral acids, and metallic salts were discovered. This lead to another, third, kind of alchemy, i.e., iatrochemy that was born as a fusion of material alchemy and spiritual alchemy, and attempted to explain chemical processes in the living body by using the medicines prepared by material alchemy. Thus, there are three kinds of alchemies:

IIa   Material (Exoteric) alchemy
IIb   Spiritual (Esoteric) alchemy
IIc   Iatrochemy or Medicinal alchemy.

IIa   Material (Exoteric) Alchemy

The first metal known to man was probably gold (Au) due to its color and luster. Gold ornaments are found dating back to Neolithic period. The next metal known was most likely copper (Cu). Nonetheless, the earliest metallurgical knowledge appears during First Dynasty in Egypt (3400 BC), and Mesopotamia. Even though, Egypt and Mesopotamia each claim the origin of working with metals, it appears that another culture that existed in Indus Valley at Harappa and Mohenjo-daro in India (now in Pakistan) which had such prior knowledge as mentioned in Vedas (about 1400 BC, see Pre-Alchemy section).

Greek science usually considered as being the basis for western science due to the fact that there are more source materials available from Greek civilization than from any other civilization. During Hellenic ear (550 – 320 BC), Greek philosophy, science, literature, and medicine dominated the Mediterranean world. Available resources suggest that their knowledge was not original as one thinks, but was derived from the technological knowledge of Indians, Chinese, Mesopotamians, and Egyptians. However, they systematized and organized into natural philosophy (philosophy and science).

Greeks were the great observers of nature and were making logical deductions to explain natural phenomena of the physical world based on their intuition rather than on myths. This is an important step in two ways;(a) got rid off old blind beliefs, and (b) pave the way to new way of thinking. Thales (640 – 546 BC) assumed that all matter was formed by water; Anaximenes(540 – 500 BC) of air; Herakleitos (536 – 470 BC) of fire. Empedokles (490 – 430 BC) introduced the idea of four elements; earth, water, air, and fire. and two forces -attractive and repulsive, which joined and separated the four elements.

Aristotle (384 –322 BC) was one of the most influential Greek philosophers who believed that all substances were created by a primary matter (*hule*), which evolves in different forms or qualities (*eidos*), just like a potter can make different kinds of pots from the same clay. He took the fundamental properties of matter, such as hotness, coldness, moistness, and dryness, and combined them in pairs to obtain Empedokles's four elements, earth, water, air, and fire (four elemental system); earth was cold and dry, water was cold and moist, air was moist and hot, and fire was hot and dry. Thus he associated two qualities with each element- duality principle. This led to the idea that one element can be transformed into another by changing the quality in that element. For example, earth can be changed into fire by replacing or changing cold into hot. Thus the idea of transmutation of elements arose.

Greek four elemental theory might have had its concepts also originated in India. The Hindu Sankya system of philosophy was supposed to be invented by Kapila about 500 BC. According to this system, knowledge of an object cannot be complete without the knowledge of its components. It defined nature (*prakriti*) as a prime matter and invoked five subtle elements (note: subtle elements are equivalent to Greek fundamental properties of matter) (*tanmatras*): sound (*shabda*), touch (*sparsha*), form (*rupa*), taste (*rasa*), and smell or odor (*gandha*), from which five course elements (*mahabhutas*): earth (*prithvi*), air (*vayu*), water(*apas*), fire(*tejas*), and ether or space (*akasha*) were created that contain from one to five of the subtle elements - five elemental system. The modification of this five elemental system in Buddhism led to six elements: earth, air, water, fire, ether, and consciousness (six elemental system).

In summary,
Greek elements (four elemental system): earth, water, air, and fire
Hindu elements (five elemental system): earth, water, air, fire, and ether/space
Buddhist elements (six elemental system): earth, water, air, fire, ether/space, and consciousness.

Aristotle was the first one to mention mercury (Hg) as " silver water." He knew a primitive method of distillation: vapor from seawater evaporated in a vessel condenses on the cool lid in the form of fresh water; and wrongly, wine could give water in a similar manner. He distinguished between chemical changes, mixtures, and solutions, was able to classify many chemical processes, proposed that metals mostly consist of water, and identified the homogeneous parts of plants and animals, thus constituting a primitive organic chemistry.

Another notable Greek theory that is important in chemistry is the atomic theory of Leucippos(c. 400 BC) and Democritus (460- c. 370 BC), and later modified by Epicourso (341- 270 BC) and Lucretius (C.95-55 BC). According to this theory, there exists only one substance called prime matter, which consists of indivisible atoms (*atomos*) of variable shapes and sizes executing random motions. Properties of material in turn dependent upon sizes and shapes of atoms.

The idea behind Greek atomic theory, however, seems to have also originated in India as Indian atomization theory predates Greek atomic theory. Atomic theory is described in Hindu *Vaisheshika* system expounded by Kanada(500 BC) and later developed in Buddhism and Jainism. According to this system, all creations are based on nine substances: earth (*prithvi*), water (apas), fire (*tejas*), air (*vayu*), space or ether (akasha), time (*kala*), direction in space (*dik*), soul (atman), and mind (*manas*). The first four substances are distinguished from one another by the special qualities of their atoms (*paramanus*). There are four such basic qualities: odor (gandha), taste (*rasa*), form (*rupa*), and touch (*sparsha*), thus radically differing from the Greek atoms. The earth possesses all four of these qualities: water possesses odor, form, and touch; fire possesses form and touch; and air possesses touch only.

The city of Alexandria in Egypt, which was founded by Alexander the Great at the mouth of Nile in 331 BC, became an intellectual center for ancient chemistry. At this place, two streams of knowledge, one ancient Egyptian Industrial art of dyeing, metallurgy, and glass-making, and the other, philosophical speculations of Greeks, which had undergone many changes and had assimilated elements of eastern mysticism, were fused together into a new kind of science. Practical artisans thought that metals in the earth tend to become perfect over the years and gradually change into gold. Hence they thought that they should be able to carry out the same process more quickly in their workshops, i.e., artificially transforming common metals into gold. This idea dominated the minds of many philosophers and artisans alike, and as a result of these, large treatises were written on the art of transformation. No one ever succeeded in making true-gold not until 1941 when Sherr, Bainbridge, and Anderson produced artificial gold from mercury making alchemists' dream come true(Partington p .379), but a number of useful chemical processes were discovered in the course of pursuit of making gold.

Greek doctrine and writings were flourished during Hellenistic era. At the end of the Roman Empire (c. 500 BC) also came the end of Greek philosophy. Greek writings were less studied in Western Europe and completely neglected in eastern Mediterranean. However, the revival of Greek philosophy got the boost in 6th century. Nestrorian Christians, a dissident sect separated from the main body of the church of Alexandria and whose language was Syriac (i.e. Theological Aramiac) spread their influence throughout Asia Minor. They established an academy at Edessa in Mesopotamia and translated large number of Greek philosophy and medical writings into Syriac.

In the 7th and 8th centuries, Arab invaders spread Islamic culture over Egypt, Persia, Syria, North Africa, and Spain. The newly created city of Baghdad, formed by Abbasid Caliphate in 762 became the center for science and learning. The Syriac translations of Greek works were again translated, this time into Arabic. Greek and other ancient doctrines once again flourished in Arabic world.

The Arabs adopted both natural Greek philosophy and alchemy as modified by Alexandrians. In addition, they were also influenced by the ideas and practices of Indian and Chinese technologists and alchemists. A specific agent known as **philosopher's stone,** was thought to facilitate the transmutation process, and hence became the main object of alchemists. They were also in contact with Indians and Chinese for invoking the idea that gold can be used as a medicine for mortality (**elixir of life**). The Arabic alchemists now had added incentive to study the chemical processes not only for the sake of becoming wealthier but also healthier.

The most famous of the Arabic alchemists were: Jabir ibn-Hayyan (about 720-813), who lived in Baghdad; al-Razi (or Rhazes, 866-925), a Persian who was an excellent physician; ibn-Sina (or Avicenna, 890-1036), a critic of alchemy; and al-Kindi (c.8000-873), who was also a critic of alchemy. Among them, Jabir and Rhazes were much revered later in the Latin west.

The idea that metals were composed of mercury and sulfur was introduced. Arabic chemistry was largely a continuation of old knowledge of old Egypt.

In Western Europe, alchemy and the Greek treatises on "divine art" were virtually unknown until the 11th century. At this time, Latin translation of Arabic works, mostly in Spain, were produced. Thus, the knowledge of Greek science, once more translated, this time into Latin, the language of European scholars. Due to translation of Arabic works into Latin was once erroneously thought that Arabs were the real inventors of chemistry.

The earliest translation on alchemy was that of Robert Chester (1140 AD). In addition, Michael Scot (1217 AD) also wrote on alchemy, but there have been some questions about the legitimacy of attributing *De Alchimia* to him.

Albertus Magnus (1193-1280) considered that alchemy was a fake science, composed about thirty-eight volumes works dealing theology, physics, and natural history, and recognized the fraud alchemists and the inconsistencies in Arab works. His *De Mineralibus* includes sections on alchemy and chemistry.

Vincent Beauvais (1190-1264), a Dominican, wrote encyclopedia, *Speculum Naturale*, mostly based on Latin translation of Arab literature that also included a section on alchemy. Another Dominican Thomas Aquinas (1225-1274), a pupil of Magnus, believed in alchemy. Roger Bacon (1214-1292) also trusted in alchemy, but was disgraced for criticizing members of his Order, and famous Dominicans like Albertus Magnus and Thomas Aquines. He classified alchemy into separate parts; (a) speculative - concerning with products of the elements, metals, salts, minerals, etc., and (b) operative - teaching methods about how to make things more effectively including gold. He described how to be more efficient by art and how to make use of medicines derived from chemical processes like distillation, sublimation, etc. He wrote a section on alchemy in *Opus Tertium*, emphasized that medicine should seek the remedies provided by chemistry, and made several alchemical works including many chemical experiments.

Arnald of Villanova (1240-1311) was a alchemist as well iatrochemist, who described the distillation of spirit from wine, said to have been produced gold artificially, and have written two alchemical books, *Rosarium Philosophorum* and *Flos Florum*.

Raymund Lully (1232-1315) was a missionary and was credited with alchemical works, the *de Secretis Naturae* and the *Testmentum*, in which he described the preparation of nearly anhydrous alcohol by dehydration using salt peter (potassium carbonate), how to prepare nitric acid and aqua regia (a mixture of nitric acid and hydrochloric acid in the ratio of 1:3 that dissolves gold and platinum), and said to have performed transmutation.

IIb   Spiritual (Esoteric) Alchemy

Spiritual alchemy was concerned with the spiritual world rather than physical world, existed well before the exoteric alchemy, and contributed the most to alchemical science and eventually became chemistry. Its aim was to understand God and find salvation by most often combined with religious rituals.

The *Vedas*, the oldest Hindu writings in India, contain some references to a connection between gold and long life. It is possible, however, that the knowledge of medicine and immortality originated in India and/or China. In both cultures, gold making was the least concern and medicine to purify the body leading towards a long life was the most concern. The Indian elixirs were mineral remedies, such as gold, for specific diseases. Among all the minerals the gold was chosen for two reasons: (a) to prolong the life, and (b) to improve the skin complexion as it has highly attractive color. The earliest Indian medical works are treatises attributed to *Susruta* (200 AD) and *Caraka* (c. 100 AD), and Bower manuscript (4th century AD). They all contain the description of metallic compounds. The *Arthasastra* (4th century BC) details mining, metallurgy, medicine, pyrotechnics, poison, fermented liquors, and sugar. *Rasaratnasamuchaya* (1200 AD) describes the preparation of zinc by distillation.

The knowledge of mercury dates back to 8th century AD, when Buddhism changed to the *Tantric* form. According to mercury system, salvation can be achieved by ingest of mercury. Early alchemists were Nagarjun (700 –850 AD), Patanjali, Narahari, Yasodhara, Krishna, and Vagbhata. The Rasarnva (c.1200 AD) mentions mineral acids. Mercurial predictions and description of chemical processes are found in *Sarngadhara* (13 century AD). Indian favored mineral methods since they always described sublimation, calcinations, and analysis.

Like Indian alchemy, Chinese alchemy was also concerned with medicines of immortality. Alchemy in China arose almost at the same time as the Taoism. Lao Tzu (c. 350 BC) introduced Taoism- a philosophy-that greatly modified by his later disciples. According to Taosim, the life can be regulated and could be prolonged.

Under Lu Nan or Huai-na Tzu, a prince of the court of Emperor Wu-ti (140-86 BC), Toaism degenerated into occultism. Frequently, elixir of life was prepared by magicians and preparers, which became one of the main objectives of Taoism. It would enable the possessor to prevent death, rise to heaven, assume other forms, and generally perform miracles.

The oldest Chinese text dealing with alchemy was ascribed to Wei Po Yang (c. 140 AD), which describes about *yin* and *yang*, *tao*, and some chemical operations, like, crystallization.

The most well known Chinese alchemist was Taoist, Ko Hong (4th century AD), who produced treatises on alchemy. His philosophy appears to have a close resemblance to Hindu Yoga system of India. He was mainly occupied with elixir of long life. The body can be prevented from decay by taking the substance *chin tan* and gold. When cinnabar(*tan shad*) is burnt it produces mercury, which can then be converted back into cinnabar by further changes. Ko Hung thought that if *chin tan* were placed over a hot fire, gold would be produced instantaneously; thus the production of gold meant the sign of completion of the elixir.

IIc   Iatrochemy or Medicinal Alchemy

Even though, Paracelsus (1493-1541) was generally regarded as the father of iatrochemy, long before him, Roger Bacon (1214-1292) and Arnald Villanova (1240-1311) had emphasized the use of chemicals prepared by alchemy in the service of medicine, and hence can be considered as first iatrochemists.

Paracelsus was born in 1493 in Switzerland in a place called Einsiedelu, obtained M.D. degree from Farrera in Italy, and was a professor of medicine at Basel from 1527 to 1528. Even though, Paracelsus spent all his lifetime in disgrace, he was a skilled physician and good surgeon. He advocated metallic, opium, and mercury remedies, believed in astrology and suggested that digestion was caused by spiritual called "Archeus" in the stomach, used extracts of plants, tinctures, essences, etc. to cure diseases, and firmly believed that morbid deposits formed in the body leads to diseases as tarter deposits from wine on long standing – theory of tarter.

Paracelsus proposed three principles (*tria prima*), sulfur, mercury, and salt, which he compared with soul, spirit, and body, also first time used the name alcohol for spirit of wine, and was the first to mention zinc (he called it " bastard metal"). His followers like Oswld Croll (wrote a book, *Basilica Chymica*) and Gerhard Dorn added more new mineral preparations, a new series of chemical remedies, described silver chloride as *luna cornea* (horn silver), and Hadrian Mynsicht described potassium antimony tartrate, $K (SbO)C_4H_4O_6$ (tarter emetic)

Van Helmont (Johann Baptista van Helmont) (1579-1644) obtained his M.D. degree at Louvain (Belgium) in 1609 and was highly respected and influential iatrochemist. His works were published in 1648 as *Ortus medicinae*, and an English translation, *Oriatrike* or *Physick Refined* in 1662. He represented the transition from alchemy to chemistry. He believed in alchemy but differed from Paracelsus in that philosopher's stone was also the elixir of life.

An important contribution of van Helmont was quantitative nature of his chemical works; he used balances, clearly expressed the "law of indestructibility of matter," and explained that metals when dissolved in acid are not destroyed, but can be recovered. He proposed "nothing is made out of nothing" and hence the weight is made of another body of equal weight. He realized, contrary to Paracelsus, that no transmutation takes place when one metal precipitates another from a solution of salt.

He was very familiar with preparation of sulfuric acid, nitric acid, aqua regia, and hydrochloric acid. By his experiments, he proved that vacuum is possible, which Aristotle thought impossible. He invented name *gas* and labeled "carbon dioxide" a *gas sylvestre*. He was proud to say that he was the inventor of gas and proposed that gas is made up of invisible atoms, coalesce and condense into liquid droplets upon intense cold. He made a distinction between gas, air, and condensed vapor. He described numerous gases like carbon dioxide, carbon monoxide, chlorine, hydrogen, methane, sulfur dioxide, and ethereal or vital.

He explained the process and purpose of respiration. He rejected the teachings of theory of five elements and three principles by Paracelsus and the Aristotle's theory of primary matter. He asserted that the true elements are air and water as neither can be converted into the other, as the elements should be. According to Boerhaave, van Helmomont's work on urinary calculi (*De lithiasi*), which also contains large number of chemical experiments, was incomparable and well written.

He fairly explained the formation of tarter in wine-casks. He observed the formation of white precipitate by mixing spirit of urine (ammonium carbonate, $(NH_4)_2CO_3$) with spirit of wine. In addition, he isolated two forms of salts from the urine, one of them was a common salt (table salt - sodium chloride, NaCl), which he asserted came from food intake, and the other of different crystalline nature. Although, Van Helmont's ideas on ferments were crude, they were in the right direction and resemble the modern theory of enzymes. He identified and distinguished six ferment processes (digestion processes) of the food when it passes through the body: the first digestion in the stomach and spleen by acid liquor; the second in duodenum by gall of the gall-bladder; the third in the mesentery; the fourth in the heart, where vital spirit is added to make red blood to become more yellowish; and the sixth when the nutrition principles are elaborated by separate ferments.

Franciscus Sylvius de le Boe (1614-1677) was a professor of medicine in Leydan, who built the first university chemical laboratory, and was mainly a theoretician.

Georg Agricolla (1494-1555), a German physician, wrote on metallurgy and mineralogy, the most famous being *De Re Metallica* published in 1556, was the first treatise on applied chemistry. He also mentioned the element bismuth (Bi).

Basil Valentine (1600?) described the preparation of many compounds of antimony (Sb), and also mineral acids like sulfuric acid (H_2SO_4) in *Triumphal Chariot of Antimony*, first published in German in 1604. There are some doubts as to the authorship of this book. It is conjectured that most likely Tholde wrote this book under Valentine's name.

Andreas Libavius (1504 – 1616), a German schoolmaster, wrote the first textbook of chemistry, *Alchemia* in 1597. He discovered stannous chloride $(SnCl_2)$ and many qualitative analysis.

Johann Rudolph Glauber (1604-1760), who was born in Bavaria and considered to be a practical chemist, invented new furnaces for distillation of salts, obtained acid and spirit by distilling the wood in closed ovens. He fairly knew that salt consisted of acid and base, and had the right idea on affinity. He prepared antimony and arsenic chlorides.

Nicolas Lemery (1645-1715) was a French chemist who wrote a text-book, *Cours de Chymie* in 1675, classified substances into three groups – minerals, vegetables, and animals based on the kingdoms of nature. He also worked on corpuscular theory- properties of substances mainly depend upon the shapes of their particles: Acids prick the tongue because they contain spiky particles and the salts of acids form sharp crystals; metals dissolve in acids because the sharp points of acids tear apart the metal particles.

Otto Tachenius, a German who lived for some time in Venice wrote a book, *Hippocrates Chimicus* (1666), which gave a clear definition of salts – compounds of acids and alkali, and studied several wet reactions and devised rudimentary system of qualitative analysis.

Johann Kunkel (1630-1703) was a skilled practical chemist, wrote a famous book, *Laboratorium Chymicum* (published in 1716), discovered phosphorus (P) independently and described the preparation of gold ruby glass, and wrote treatise on glass manufacturing, *Ars Vitraria* (1679) in German.

III   Post-Alchemy (Mid 17th Century – Today)

The transition from alchemy to the start of modern chemistry took place in the middle of 17th century through two kinds of chemical changes, namely, combustion and calcinations of metals, which were studied side by side and concerted efforts were made to describe them together. Alchemists, with their many types of furnaces gave a greater importance to the effects of heat on substances. The metals, except, silver and gold, were found to transform into dross (this was called calx meaning lime from Latin) when heated in open crucibles. Since the calx was heavier than the metal, as it was known in the sixteenth century, various explanations were put forward: some kind of "soul" escaped from the metal or that fire possessed weight and was absorbed by the metal or that some kind of acid was absorbed by the metal - a crude but reform thinking compared to previous thinking.

IIIa   Birth of Chemistry

During the mid 17th century, the weakness in alchemical thinking was beginning to emerge. New ways of thinking and explaining chemical processes were introduced. As results, new concepts and theories were surfaced which were later to serve as the foundation for start of early chemistry

IIIa.a   Early Knowledge of Chemistry

Few investigations of carbon compounds here and there go beyond 19th century: alcohol was described in 12th century manuscript; ether was described in a work of Valerius Cordus around 1544; Boyle described the distillate from boxwood containing methyl alcohol and acetone in his *Sceptical Chymist* (1661); Balise de Bigenere (1522-1596), whose work was published in 1618 gave an account of benzoic acid; and Scheele (1742-1786) discovered several organic acids (tartaric, mucic, lactic, uric, prussic, oxalic, citric, malic, gallic, and pyrogallic in fruits and plants), glycerin, hydrofluoric acid, several esters, aldehydes, and casein.

In the early part of the 19th century, more emphasis was placed on the chemistry of metals and common elements like nitrogen, phosphorus, and sulfur rather than on carbon compounds. Chemistry was divided into two groups, Animal chemistry and Vegetable chemistry based on their source of origin: in animal chemistry, urine, saliva, blood, urea, gelatin, albumin, fibrin, and similar natured substances were described mostly medical point of view and in vegetable chemistry, acids, sugar, gum, indigo, tannin, camphor, and India- rubber were described.

All the investigators of 19th century were both inorganic and organic chemists who realized that the above mention substances contain carbon and hydrogen, some times oxygen, nitrogen, and sulfur. Their analyses were in a very primitive conditions as many substances could not be crystallized or purified.

As late as 1835, Wohler writing to Berzelius expressed his feelings about organic chemistry as "Organic chemistry appears to me like a primeval forest of the tropics, full of most remarkable things (Partington p.216)".

IIIa.b The Oxford Trios

Further investigations on combustions and calcinations were performed by the so called "Oxford chemists" – Robert Boyle (1627-1691), Robert Hooke (1635-1702), and John Mayow (1641-1679), among whom, the Boyle, who also believed in alchemy, has been called as the founder of modern chemistry for (a) realizing that chemistry is worthy of study, not merely as an aid to medicine as iatrochemists thought, (b) introducing rigorous experimental techniques, and (c) giving clear definition of an element; showed by his own experiments that four elements of Aristotle (air, water, earth, and fire) and three principles of alchemists (mercury, sulfur, and salt) do not deserve to be called either elements or principles as none of them could be extracted from bodies. In addition, he did numerous quantitative experiments; among them the most notable was the quantitative relationship between the volume and the pressure of the gas, which was transformed into a gas law known as the Boyle's law (1662).

Hooke is well known for his discovery of Hooke's law (1660)- the extension of a single or any stretchable object is directly proportional to the force acting on it. He made further experiments on combustion and proposed that air contains a substance that acts as a solvent for combustibles.

Mayow, who was a physician, had a clear idea about chemical affinity and tried to explain the combustion which he thought was nothing but the collision between nitro-aerial particles in air with sulfurous particles in combustibles. He also explained the increase in weight of metals in calcinations is due to absorbing of nitro-aerial particles from the air by the metals. He suggested that combustion and respiration in animals are analogous processes. He proved that arterial blood in vacuum gives off gas, and heat is generated in animal muscles.

IIIa.c   The Theory of Phlogiston

The idea behind the theory of Phlogiston was developed by Johann Joachim Becher (1635- 1682). In 1669, he published a book titled, *Physica Subterranea*, in which he described the constituents of bodies are air, water, and three earths - one of which is "inflammable", the second mercurial, and the third vitreous (fusible) - which corresponded with the sulfur, mercury, and salt of the alchemists. He noted that when combustion takes place, an inflammable material leaves the burning body and enters the air.

Goerg Earnst Stahal (B 1660) improved Becher's views and popularized it by naming Becher's inflammable as the Greek word "Phlogiston." He believed that all combustible substances and metals contain phlogiston that can be expelled to the air when these substances are burned. Furthermore, he emphasized that burnt substances may be restored to original substances by supplying the phlogiston from a material rich in it, such as, wax, oil, charcoal or soot, thus completing the cyclic process. For example, zinc metal when heated to redness burns with brilliant flame by making phlogiston to escape. When white residue (calx of zinc) is heated to redness with charcoal (rich in phlogiston), the original zinc was recovered. Hence, calx of zinc + phlogiston = zinc. Stahal did not have any idea about the specific nature of the phlogiston, but for him, it was some kind of material; some times invisible particle, some times a dry earthy substance like soot, some times the principle of fire or some other times a fatty principle like oil, fats, resins and sulfur.

The concept behind the phlogiston theory is very similar to the concept of reincarnation dogma (theory) in Hindu religion. In Hinduism, it is believed that when a living body dies, the soul (like phlogiston has no specificity) leaves the body and goes to heaven. Later, some point in time, the soul returns to earth and enters into some other body making it animate. This cycle goes on and on.

Stahal in proposing his theory, paid no attention to quantitative aspects of chemical changes, known facts about the gases, existing atomic theory, and most of all, the three alchemical principles (sulfur, mercury, and salt). This prompted numerous modifications of his theory. Cavendish (1731-1810), Kirwan (1735-1812), and Priestley (1733-1804) identified phlogiston with hydrogen (inflammable air). Baume (1777) suggested that phlogiston is composed of an earth and the matter of fire in various ratios, and metals on calcinations absorb pure fire by losing phlogiston. Macquer (1779) assigned the phlogiston to the matter of light.

After Lavoisier proposed Oxygen theory of combustion, an attempt was made to retain some features of the phlogiston theory, notably, by Lubbock (1784), Gadolin (1788), and Richter (1791), who argued that a combustible substance might contain a base-material united with phlogiston and oxygen gas (substrate-material) combined with heat. When the combustion takes place, both materials unite by attraction and at the same time phlogiston and heat unite by affinity to give off light and fire.

Lemonsov (1745) explained the increase in weight of metals upon calcinations is due to air fixed by the metals disproving Boyle's hypothesis that was due to fixation of fire particles. On the other hand, Laurent Beraut assumed that air contains certain foreign particles, which get separated by the action and combine with metal during the calcinations leading towards an increase in weight.

Notable chemists of phlogiston's period were: Wlihelm Homberg (1652-1715, French), Friedrich Hoffmann (1660-1742, German), Etienne Francois Geoffroy (1672-1731, French), Casper Neumann

(1683-1737, German), Johann Heinrich Pott (1689-1777, German), Guillaume Francois Rouelle (1703-17770, French), Andreas Sigismund Marggral (1709-1782, German), Pierre Joseph Macquer (1718-1784, French), Joseph Black (1728-1799, English), Henry Cavendish (1731-1810, English), Joseph Priestley (1733-1804, English), Richard Kirwan (1735-1812, English), Torbern Bergman(1735-1784, Swedish), Carl Wilhelm Scheele (1742-1786, Swedish), Johann Gottlieb Gahn(1745-1818, Swedish), and William Higgins (1763-1825, English).

Even though, the theory of phlogiston had a great success in coordinating a large volume of facts into system, it prevented many best investigators from the correct interpretation of facts, and thereby, hindering the progress of chemistry.

IIIb   Foundation of Modern Chemistry

Antoine Laurent Lavoisier (1743-1794) has been credited as the father of modern chemistry. He had an excellent education in chemistry, physics, and mathematics. He gave the greater importance to the use of balances and quantitative methods, and correctly interpreted many previous chemical processes. He stated the law of indestructibility of matter (1789) ("law of conservation of mass" in its current version) contrary to previous belief. He proved that water cannot be converted into earth, correctly interpreted the increase in weight in calcination or combustion due to absorption of equal amount of air, proved that air consists of a special kind of gas and named it as oxygen that is being absorbed in the combustion, showed that combustion of non-metals produces acids, and oxidation of metals produces bases ("calces"), replaced phlogiston theory with anti-phlogiston theory stating that loss of phlogiston is nothing more than absorption of oxygen and vice versa, analyzed combustion of organic substances in oxygen, correctly explained the composition of water based on Cavendish's experiments, devised new chemical nomenclature in collaboration with A.F.deFourcroy(1755-1809) and C.L. Berthollet (1748-1822), and Guyton de Morreau (1737-1816).

With such new wave of correct interpretations, the theory of phlogiston rapidly disappeared after 1785, but clung on to very few conservative chemists like Priestley, Macquer, etc.

In a further quest to overhaul the old chemistry, first came the "law of constant proportion" proposed by French chemist Joseph Louis Proust in 1797 followed by Berthollet's " law of definite proportions". Then came the "law of multiple proportion " of W.H. Wollaston in 1808.

The "equivalent" was invented by Cavendish in 1766 who stated that different weights of different bases are required to neutralize the identical weights of a given acid. Jeremias Benjamin Richter (1762-1807) introduced the term "stoichiometry" and defined the equivalents and determined the equivalents of various substances including metals, acids, and bases. Further, E.G.Fischr (1754-1831) published Richter's numerous tables into a single table of equivalent weights. Then came the Dalton's atomic theory that revolutionized the modern chemistry.

IIIb.a Electrochemical or Dualistic Theory
Humphry Davy (1778-1829), a British chemist, poet, and a philosopher, investigated properties of nitrous oxide including its physiological action. He isolated the metals of the alkalis, alkaline earths, and boron, attempted to isolate fluorine, worked on chlorine oxides and iodine oxides, conducted research on the nature of flame and lamps for mines, worked on effect of electricity on chemical reactions, and put forward somewhat imprecise electrochemical theory.

Michael Faraday (1791-1876), a British chemist, is credited for the discovery of laws of electrolysis. In addition, he discovered liquefaction of chlorine, butylenes, benzene gases, and two chloride of carbon.

Joseph Louis Gay-Lussac (1778-1850), a French chemist, discovered the law of gaseous volumes and the law governing the effect of heat on volume expansion. He carried out volumetric analysis of silver titration, chlorimetry, and acidimetry. He was able to make a distinction between oxides of nitrogen, NO_2 and N_2O_3 , isolated cyanogens (an organic radical), and studied cyanogen compounds and iodine.

Louis Jacques Thenard (1777-1857), a French chemist, discovered hydrogen peroxide and investigated oxides of phosphorus, metals, organic substances like ethers, bile, and sebacic acid. In collaboration with Gay-Lussac, he worked on alkali metals and chlorine, and proposed improved method of preparation of alkali metals. Together, they discovered potassium and sodium peroxides, and potashamide, and showed that hydrogen and oxygen as an integral part of caustic potash and soda.

Jon Jacob Berzelius (1779-1848), a Swedish chemist, is considered to be great influence on the development of contemporary chemistry. His precise quantitative experiments showed that the atomic theory and the laws of combination could be applied in both inorganic and organic chemistry. He classified minerals based on their chemical composition, proposed modern chemical symbols, and developed an electrochemical theory, independent of Davy, with different point of view. His theory was dualism: according Lavoisier, the acid was a compound of radical with oxygen; according to Davy, the base was a compound of metal with oxygen; but according to Berzelius, the salt was a compound of an acid and base. Berzelius extended this dualism concept to electrochemical reactions and stated that when electricity passed through a salt solution, bases (he identified them as electropositive substances) migrate towards negative poles, and the acids (he identified them as electronegative substances) migrate towards positive poles. He created tables of atomic weights, discovered ceria, selenium, and thorium, isolated zirconium, titanium, and silicon, and made thorough studies on compounds of tellurium, of rare metals, and of sulfa-salts. In his laboratory, the compounds of lithium were discovered by Arfvedson, although the metal itself was discovered by Davy. He was instrumental in using filter-papers, water baths, wash bottles, desiccators, rubber-tubing, blow-pipe analysis, thereby improving analytical methods as well as organic combustions,

investigated pyruvic and sarcolactic acids, and other organic compounds, was the inventor of the term "isomerism" based on tartaric and
racemic acids (both have same composition), and suggested the term
"catalysis", and introduced the name "halogens." His annual reports and
a textbook were simply authoritative.

Eilhard Mitscherlich (1794-1863), a German chemist, developed
"contact theory" of etherification and fermentation. He not only recognized the catalysis but also labeled it as "contact action." He discovered
monoclinic sulfur, selenic acid, benzene sulfonic acid, and nitrobenzene, obtained benzene from benzoic acid, and invented the process of
isomorphism, and polymorphism.

Pierre Louis Dulong (1785-1838), a French chemist, invented nitrogen trichloride, studied gravimetric composition of water, lower oxyacids of phosphorus, and oxides of nitrogen. He also measured
refractive indices and specific heats of gases, and proposed a " hydrogen
theory" of acids and bases similar to the one suggested by Davy. In collaboration with Petit, he formulated the "law of constancy of heat",
which is known as "Dulong-Petit law"; product of specific heats (in J/g
^{0}C) and molar masses of metals is approximately 25.0. This law was
used by him and later by Cannizzaro in determining atomic weights.

William Prout (1785-1850), a British medical doctor, formulated the
concept that hydrogen is the fundamental element from which other
elements are formed by condensation. Based on this, he proposed that
all atomic weights are multiples of hydrogen. He discovered uric acid,
murexide, the procedure for urine analysis, and organic combustion
with oxygen.

Amedeo Avogadro (1776-1856), a Italian physicist, proposed an
important hypothesis, which states that equal volumes of gases contain
identical number of particles (atoms or molecules) at the same temperature and pressure. Now this number is known as Avogadro's number,
which is 6.022×10^{23}. He showed that molecules of elementary gases
like hydrogen, oxygen, and chlorine consist of two atoms not of single

atom as Dalton thought. Thus, he was able to reconcile Dalton's atomic theory with Gay-Lussac's law of combining volumes.

Thomas Thomson (1773-1852), a British chemist, who published an account of Dalton's atomic theory in the form of *system of chemistry* in 1807. He discovered chromyl chloride and sulfur chloride, and pointed out multiple proportions of oxalates. He was greatly influenced by Prout's hypothesis and also determined many atomic weights.

It should be noted, however, that good early atomic weight determinations were performed by two British chemists, Edward Turner in 1833 and F. Penny in 1839.

"Group tables" for qualitative analysis based on Berzelius's method were produced by two German chemists; Heinrich Rose (1795-1864) in the form *of Handbuch der analytischen chemie*, and Carl Remigius Fresenius (1818-1897) in *Treatises on qualitative and quantitative analysis*.

Antoine Jerome Balard (1802-1876), a French chemist, invented bromine in 1826. In addition, he also invented hypochlorous acid, and chlorine monoxide in 1834.

Leopold Gmelin (1788-1853), a German chemist, wrote a large *Handbook of chemistry,* introduced the names 'ketone' and 'ester,' invented potassium ferricyanide, taurine, croconic acid, rhodizonic acid, haematin, and pancreatin with Tiedemann.

IIIb.b   The Radical Theory

The radical in the modern sense first used by de Morveau in 1787. The radical in a simple compound consists of carbon and hydrogen in addition to different proportion of oxygen, e.g. oxalic acid was the higher and the sugar the lower oxide of hydrocarbon radical. The oils, on the other hand, being free from oxygen, might be considered as free radicals. Gay-Lussac clearly demonstrated the existence of organic radical, cyanogens in 1815. Berzelius while extending his dualism theory to organic compounds considered that all organic substances are oxides of compound radicals. The radical of animal substance generally contain carbon, hydrogen, and nitrogen, and those of vegetable substances contain carbon and hydrogen.

Michel Eugene Chevreul (1786-1889), a French chemist, must be considered as one of the founders of modern organic chemistry. He studied the composition of oils and fats, and vegetable colors, clearly explained the saponification reaction, and investigated the analysis of organic compounds. He often used the melting point as a method of identification.

Justin von Liebig (1803-1873), a German chemist, was one of the outstanding chemists of the first half of 19th century, who was a clear thinker and excelled both in theory and experiments. He carried out impressive amount of experimental work in the field of organic chemistry. He invented hippuric acid, synthesized chloroform and chloral, collaborated with Wohler on benzoyl compounds and recognized benzoyl radical, and was instrumental in adopting ethyl radical (this was also simultaneously suggested by Kane and Berzelius). He revived Davy's and Dulong's "hydrogen theory" of acids, extended Graham's research to a number of organic acids and formulated polybasic acids, suggested "vibrating" theory of decay and fermentation, studied aminoacids, amides from animal products, creatine, and creatinine. He also formulated theories of physiological interest, and introduced mineral manures and meat extract. He discovered a method for silver mirroring and devised the cyanide process for separating cobalt from nickel. He edited *Annalen der Pharmacie* (1832- 1839), which continued as *Annalen der Chemie und Pharmacie* from 1840, and as *Annalen der Chemie* from 1874.

Friedich Wohler (1800-1882), a German chemist, made significant contributions to organic as well as inorganic chemistry. In organic chemistry, in collaboration with Liebig suggested that oil of bitter almonds (benzaldehyde), benzoic acid, benzamide, and benzoyl chloride all contain group of atoms (he wrote it as $C_{14}H_{10}O_2$) and called it benzoyl. He also investigated mellitic acid and invented parabanic acid and amygdalin. He discovered quinhydrone and hydroquinine, and tellurium methyl. He obtained urea from ammonium cyanate and also by the action of light on a mixture of chlorine and carbon monoxide and mistakably called it as phosgene. In inorganic chemistry, he isolated aluminum, beryllium, crystalline boron and silicon, prepared silicon nitride, and with Buff discovered silicon hydride and silicon chloroform. He discovered calcium carbide and obtained acetylene from it, recognized similarity between carbon and silicon compounds. He also worked on metallic sub- oxides and peroxides, and preparation of phosphorus by modern method. He translated several editions of Berzelius's textbook into German.

Robert Wilhelm Bunsen (1811-1899), a German chemist, invented many chemical apparatus like Bunsen burner, Bunsen battery, grease-spot photometer, absorptiometer, actinometer with Roscoe, effusion apparatus, filter pump, and ice calorimeter. He worked out analytical methods like iodine titration, studied action of light on chemical reactions with Roscoe, applied spectroscope to chemistry and found that each element has a characteristic emission spectrum, and as result of which, he discovered cesium and rubidium. In addition, he investigated cacodyl compounds and recognized cacodyl radical contained in them, and isolated cacodyls $As_2(CH_3)_4$. He also studied the formation of cynamides from alkalis, carbon, and nitrogen. Bunsen's classical work on the cacodyl radical supported the radical theory extended by Liebig based on the views of Lavoisier and Berzelius.

IIIb.c   Theory of Substitution and Other Theories

Two French chemists, Dumas and Laurent, introduced a new way of looking at reactions in organic chemistry, which led to the downfall of the dualistic theory of Berzelius. The molecule as a whole was considered to be a structure, modifications of its parts leads to structurally related molecules. This new point of view was labeled as unitary theory based on the concept of substitution.

Jean Baptiste Dumas (1800-1884), a French chemist, formulated a theory of substitutions, in which he summarized as " when a substance containing hydrogen is reacted with chlorine, bromine, iodine or oxygen, it gains an atom of chlorine, bromine, iodine or half atom of oxygen for each atom of hydrogen it loses". He investigated vapor densities, and with Boullay put forward the etherian theory; alcohol, ether, and esters are derived from ethylene. He discovered oxamide, recognized methyl alcohol in wood spirit, established general series of alcohols, studied the hydrolysis of nitriles to acids and amines, invented combustion method for determining organic nitrogen, and also made accurate determination of atomic weights.

Auguste Laurent (1808-1853), a French chemist, discovered chrysene, pyrene, benzil, anthaquinone, phthalic acid, adipic and pimelic acids, isatin, and with Dumas an anthracene. He identified picric acid and devised methods for classification of organic compounds. He pointed out that the compounds still retain their essential chemical properties after substitution. He suggested ether and alcohol corresponded with K_2O and KHO (nucleus theory), and criticized the dualistic theory in his posthumous *Chemical Method*.

Charles Gerhardt (1816-1856), a French chemist, discovered quinoline, anilides, acetylchlorine, acid chlorides, and acetic anhydride. Proposed reformed system of atomic weights. The molecular weight was defined as the weight of the vapor occupying the same volume as 2.0 gram of hydrogen in vapor state. This was known as two volume atomic weights. His atomic weights were less accurate than Berzelius as he assumed the water type Me_2O for oxides of most metals, but corrected Berzelius's atomic weights of silver and alkali metals by dividing them by 2. He made a significant contribution to the Theory of Residues, recognized homologous series of compounds, and also proposed theory of four types (H_2O, NH_3, HCl, and H_2), which included some aspects of radical theory that was in agreement with Laurent's Unitary Theory. His textbook, Organic chemistry, the last volume of which listed the formulation of organic compounds with atomic weights was simply considered to be authoritative.

Both Laurent and Gerhardt were brilliant chemists of their time, but were badly treated by some their colleagues.

Auguste Cahours (1813-1891), a French chemist, invented amyl alcohol, anisole and its derivatives, methyl salicylate, use of PCl_5 in the synthesis of organic acid chlorides, tin tetraethyl, allyl alcohol, and allyl sulfonium bases. He also studied abnormal vapor densities.

Thomas Anderson (1819-1874), an English chemist, discovered collidine in bone oil, lutidine, picoline, and pyridine. His other contributions were the constitution of anthracene and piperdine, and the preparation of pure pyrrol.

August Wilhelm von Hofmann (1818- 1892), a German and British chemist, did very extensive work in organic chemistry by working in Berlin and London. He discovered phenyl isocyanate, hydrabenzene, diphenylamine, isonitriles, formaldehyde, and phenyl mustard oil. With Martius, he invented methyl and dimethyl aniline. His discovery of reaction of bromine and alkali with amides led to a general method known as " Hofmann reaction". He also worked on alkaloids and aniline dyes. He founded the German Chemical Society in 1868, and was a brilliant lecturer, who incorporated the new " two volume atomic weights" and numerous lecture experiments into his lecture.

Alexander William Williamson (1824-1904), an English chemist, put forward a dynamic view of chemical equilibrium, extended Laurent's concept of water type (alcohol and ether are analogous to water). This was an important step towards the understanding that molecules of alcohol and ether occupy the same volume in a vapor state. He recognized the acetyl radical and the structure of acetone. He discovered chlorosulfonic acid, and prepared aldehydes and ketones by distilling calcium salts of organic acids. He devised the synthesis of unsymmetrical ethers, which is known as "Williamson synthesis".

Stanislao Cannizzaro (1826-1910), an Italian chemist, carried out important research in organic chemistry and is well known for discovering the process of formation of benzoic acid and benzyl alcohol when potash reacted with benzaldehyde, which is known as "Cannizzaro's reaction". His other contributions included determination of atomic weights by the application of Avogadro's hypothesis and Dulong and Petit's law of atomic heats.

IIIb.d The Theory of Valence
The development of this theory coincided with the development of the theory of "types". But the chemists of that time were uncomfortable with the theory of types and believed that something is hidden deeper. The precise work of Kolbe and Frankland, guided by the old teachings of Berzelius, revealed this deeper meaning, thereby making the type theory to collapse.

Hermann Kolbe (1818-1884), a German chemist, who was greatly influenced by Berzelius's ideas, foreshadowed the modern structural formulas and valence, and predicted the existence of compounds like secondary and tertiary alcohols. He synthesized acetic acid, methyl sulfonic acid, and salicylic acid, studied the electrolysis of salts of fatty acids, and worked on taurine, malonic acid, and aliphatic nitro-compounds.

Edward Frankland (1825-1899), an English chemist, put forward the theory of valence based on his own investigation of organo-metallic compounds, and devised modern structural formulas. He suggested that valence of an element could vary. In addition, he discovered organo-metallic compounds of zinc, was able to prepare hydrocarbons from zinc alkyls, investigated constitution and reactions of acetoacetic ester with Duppa.

Christian Wilhelm Blomstrand (1826-1897), a German chemist, also contributed to the theory of valence by expressing varying valences of an element in different compounds, e.g., Cl 1 in HCl and 7 in $HClO_4$. He also investigated complex amine compounds of metals and proposed diazonium formula for diazo-salts independent of Strecker and Erlenmeyer.

August Kekule (1829-1896), a German chemist, extended and developed the work of Frankland, recognized the tetra valence of carbon (1858) independent of Archibald Scott Couper, and linking of carbon atoms that can explain structures of many organic compounds. He was firm believer in the constant valence of element, and invented hexagonal structure of benzene (1865). As a result, it was felt that theory of types was no longer necessary making a room for the birth of modern structural chemistry. He also carried out special researches in different branches of chemistry; discovered triphenyl, prepared acetylene by the electrolysis of fumaric acid, and recognized crotonaldehyde.

Charles Adolphe Wurtz (1817-1884), a French chemist, was the first one to teach Gerhardt's teachings in France. He discovered the amines – organic compounds containing the radical NH_2 – predicted by Leibig. His textbook had a great influence on the modern chemical theory. He discovered phosphorous oxychloride, $POCl_3$, studied the lower oxyacids of phosphorus, carried out the synthesis of amines from alkyl isocyanates, glycols and ethylene oxides, obtained hydrocarbons from alkyl halides and sodium oxide, investigated the reduction of aldehydes to alcohols by sodium as well as aldol condensation.

Marcellin Berthelot (1827-1907), a French chemist, was regarded as one of the most distinguished chemists of 19th century. His research was highly original and of fundamental in nature, may be divided into four periods: from 1850-60, he worked on the constitution and synthesis of organic acids and polyatomic aliphatic alcohols; from 1961-70, he concentrated on the synthesis of hydrocarbons- synthesis of acetylene in presence of an electric arc, its polymerization to benzene on reduction of organic compounds by hydriodic acid in a sealed tube, and on the rates of chemical reaction with Sainte-Gilles; from 1869-85, he diverted his attention to physical chemistry – carried out research on thermodynamical reactions, invented " detonation wave in explosives", and studied specific heats of gases at high temperatures; from 1885-1907, he did research on agricultural and historical chemistry – his writings on history of alchemy gave a complete and critical view of the alchemists of Alexandrian period.

IIIc   Renaissance in Chemistry

If we examine the history of chemistry more closely, we certainly notice that there was a period of time in the history – mid 19th century to early 20th century – that can be certainly labeled as the renaissance period. During this period, the discoveries in chemistry were phenomenal with further twigging of chemistry into four major branches of chemistry, which are described below.

IIIc.a   The Rise of Organic Chemistry

The development of organic chemistry took place in the later part of 19th century and the early part of 20th century. Many new physical and chemical properties of compounds were discovered coupled with synthetic organic chemistry led to the recognition of many class of compounds. New synthetic methods and analysis were introduced to carry out substitution.

Louis Pasteur (1822-1895), a French chemist, investigated optical activity of organic compounds. He observed hemihedral part of the crystals of tartaric acid and tartrates, resolved optically inactive or racemic compounds by two methods; (a) by chemical means – fractional crystallization with optically active base or acid; for e.g. sodium ammonium racemate, and (b) by bacteriology – growing a mould that selectively uses up one form. He discovered meso- and levo-tartaric acids. His researches on fermentation started the whole new science - microbiology.

Van't Hoff (1852-1911), a Danish chemist (Netherlands), independently studied the relationship between an asymmetric carbon and optical activity in 1874. He also laid the foundation for modern physical chemistry. His contribution to this discipline includes general treatment of velocity, the application of thermodynamics to chemistry, dynamic equilibrium, transition point in heterogeneous systems, affinity as measured by the diminution of electromotive force, relation between equilibrium constant, heat of reaction and temperature, and the theory of dilute solutions.

Hans Landolt (1831-1901), a German chemist, also studied the optical activity in addition to organic compounds of arsenic and antimony. He tested the empirical formula developed by John Hall Galdstone (1850-1911) and J. Dale in 1858 for the refractive power of compounds.

J.W. Bruhl (1850-1911), a German chemist, on the other hand, tested the theoretical formula on molecular refractions proposed independently by L. Lorenz and H. Lorenz with a special emphasis on the influence of molecular structure.

Emil Erlenmeyer, Sr. (1825-1909), a German chemist, synthesized tyrosine and elucidated the structure of lactones. He invented isobutyric acid, studied the constitution of lactic and hydracrylic acids, and naphthalene. In addition, he also proposed a structure notation and theory of valence.

Alexander Michailowitsch Butlerow (1828-1886), a Russian chemist, recognized the existence of dynamic isomerism, studied isomeric dibutylenes, synthesized hexoses and tertiary alcohols from acid chlorides and zinc alkyls. Johann Peter Griess (1829-1888) invented aromatic diazo-compounds. Hans von Pechmann (1850-1904), a German chemist, discovered aliphatic diazo compounds by the reaction of alkalis with nirosamines.

Carl Schorlemmer (1834-1892), a British chemist, discovered general method for conversion of secondary alcohols to primary alcohols, worked on paraffin hydrocarbons, boiling points of paraffins, studied the constitutions of aurin and suberone. He also discovered the existence of supposed free methyl (CH_3) and ethyl hydride ($C_2H_5.H$), which were considered as identical isomers with molecular formula C_2H_6. He wrote a large *Treatise on Chemistry* with Roscoe.

Rudolf Fittig (1835-1910), a German chemist, discovered the synthesis of aromatic hydrocarbons by the action of sodium on a mixture of alkyl and aryl halides – Fittig's reaction, the pinacone reaction, diphenyl, isophthalic acid, phenanthrene, coumarone, diacetyl, synthesized mesitylene, and a-naphthol, investigated complex salts like K_4Mn $(CN)_6$ and $K_3 Mn(CN)_6$, ketonic esters, piperine, and unsaturated carboxylic acids, put forward the diketone formula for quinone, described the relationship between γ-hydroxy acids and lactones.

William Henry Perkin, Sr. (1838-1907), a British chemist, was the first to synthesize aniline dye, mauve, coumarin, cinnamic acid (Perkin reaction), and glycol and tartaric acid with Duppa. He studied the relations between chemical constitution and magnetic rotatory power, and constitution of saligenin.

Johann Wislicenus (1835-1902), a German chemist, adopted and extended Van't Hoff's stereochemical theories and applied to geometrical isomerism of unsaturated compounds. His discovery of simple way to synthesize hydrazoic acid is very important. He also discovered vinyl ether and vinyl acetic acid, investigated the condensation of aldehyde ammonia with lactic acids, use of acetoacetic ester in organic synthesis and synthesis of glutaric acid, methyl β-butyl ketone with Conrad and Limpach. Along with his students, Conrad and Limpach, he worked out the details of the method of synthesis depending on the interaction of an alkyl iodide with sodium derivatives of acetoacetic ester.

Adolf Baeyer (1835-1917), a German chemist, discovered indole and the constitution and synthesis of indigo, put forward the strain theory of carbon rings, investigated purine derivatives, reaction products of benzene and its derivatives.

Terpenes were also investigated by Albin Haller (1849-1925), a French chemist, Otto Wallach (1847-1931), a German chemist, and

William Henry Perkin, Jr., a British chemist, and Gustav Komppa (b. 1867) in Helsingfors.

Arthur Hantzsch (1857-1935), a German chemist, advanced theories of indicators and of the structure of acids, applied physico-chemical methods, like electric conductivity to organic chemistry, investigated stereochemistry of nitrogen, the structure of oximes (with Werner), diazo-compounds, the structure of cyanuric acid and cyamelide, synthesized pyridine from acetoacetric ester and aldehyde ammonia, coumarone, and thiazole, isolated hyponitrous acid, and studied tautomeric behavior of phenyl nitromethane and nitro-phenols as pseudo-acids.

Ludwig Knorr (1859-1921) worked on tautomerism that led him to define a tautomeric compound as a mixture of two forms in dynamic equilibrium (allotropic mixtures) based on his own investigation of diacetyl succinic ester, acetoacetic ester, and acetyl acetone. He devised colorimetric method to determine the ratio of enol-form and keto-form in a tautomeric mixture with H. Schubert, devised a preparative method for acetonylacetone, investigated pyrazolone and isopyrazolene and their derivaties, discovered an important antipyrene drug- phenyl dimethyl pyrazolene, and aminoethyl ester with G. Meyer, worked on piperazine derivatives and alkaloids.

William Henry Perkin, Jr. (1860-1929), a British chemist, synthesized and worked on polymethylene rings that led to Baeyer's strain theory. In addition, he carried out numerous syntheses like ethyl benzoyl acetate and derivatives, anthraquinone by heating ortho-benzoylacetic acid with sulfuric acid, indene and hydrindene and derivatives, camphor and its derivatives (camphoronic acid with J.F. Thorpe), alkaloids (bererine, narcotine, cotarmine, terpineol, cryptopine, protopine), terpens (synthesis of terpinol), brazilin and haematoxylin, harmine and harmaline, and isoquinoline derivaties.

Johannes Thiele (1865 –1918), a German chemist, put forward a theory of partial valences to explain addition reactions to double bonds and constitution of benzene, discovered semicarbazide, derivatives of fulvene (a colored hydrocarbon), and investigated tetrazole derivatives, nitramide, guanidine derivatives, and unsaturated lactones.

Victor Grignard (1871-1935), a French chemist, invented the use of magnesium alkyl halides (MgRX) in place of metal alkyls in synthetic reactions, which is known as Grignard reaction.

Richard Willstatter (1872-1942), a German chemist, synthesized betaine, lecithin, ortho-quinones, quinone-imines, pyrones (chlorophyll), the coloring matters of flowers (anthocyanin, etc.), blood, etc., investigated the assimilation of carbon dioxide by plants and enzymes, alkaloids and their derivatives (tropic acid, tropine, atropine, ecgonine, and cocaine).

Paul Sabatier (1854-1941), a French chemist, introduced the method of catalytic hydrogenation and obtained pure hydrogen disulfide.

Henry Edward Armstrong (1848-1937), a British chemist, studied naphthalene derivatives, terpenes, and camphor, investigated quinonoid theory of color of dyestuffs, wrote on organic and inorganic chemistry and teaching science. He was critical of the theory of electrolytic dissociation.

William Jackson Pope (1870-1939), a British chemist, mainly worked in the field of stereochemistry and particularly on optically active substances containing carbon and other elements. He pioneered in introducing camphor sulfuric acids in resolution of optically active bases and hydroxyhydrindamine with Read in the resolution of acids. He prepared mustard gas (dichloro diethyl sulfide) with C.S. Gibson by the action of ethylene on sulfur chloride.

Gilbert Thomas Morgan (1870-1940), a British chemist, carried out investigations related to dyestuffs; coordination compounds, catalytic hydrogenation, phenol-formaldehyde condensation products, organic compounds of arsenic, antimony, selenium, and tellurium.

Arthur Lapworth (1872-1941), a British chemist, applied electronic theory of valence to organic chemistry, worked out cyanohydrin formation, mechanism of esterification, enol-keto change in the bromination of acetone, basic properties of water and alcohol, and a general theory of acids and bases that provided a ground for Lowery and Bronsted to propose their theory of acids and bases.

Jocelyn Field Thorpe (1872-1940), a British chemist, studied imino-compounds, hydrindene derivatives, glutacomic acids, tautomerism and bridged rings, camphor and terprens – synthesis of camphoric and camphoronic acids with Perkin.

Sir Robert Robinson (1886-1975), a British Chemist, worked on alkaloids and synthesis of many useful medicinal agents by making a great contribution towards pharmaceutical industries.

IIIc.b   The Rise of Physical Chemistry

The development of physical chemistry also took place in 19th century along the side of the development of organic chemistry and inorganic chemistry. Some physical chemists were also organic chemists. This field also attracted some mathematicians and physicists. Jacobus van't Hoff (1852-1911), Svante Arrhenius (1859-1927), and Wilhelm Ostwald (1853-1932) are generally considered as the founders of Physical Chemistry.

Thomas Graham (1805-1869), a British chemist, was also considered to be one of the founders of physical chemistry. He studied diffusion, effusion, and transpiration of gases and proposed that rate of diffusion of the gas is inversely proportional to the square root of the density – Graham's law of diffusion. In addition, he also investigated numerous other systems like absorption of gases and vapors by liquids, solubilities of various salts, process of supersaturation, oxidation of phosphorus, adsorption of dissolved salts on charcoal, arsenates, phosphates, and alcoholates leading to the theory of polybasic acids, theory of voltaic circle, hydrated salts and oxides, dialysis and osmosis, liquid diffusion, colloidal silica, and occlusion of hydrogen by metals.

Hermann Kopp (1917-1892), a German chemist, was also one of the founders of physical chemistry, and perhaps best known as a historian of chemistry. He tried to establish relations between the properties of substances and chemical composition by investigating atomic and molecular volumes, crystallography, boiling points, specific heats, and dissociation.

August Horstmann (1842-1929), a German chemist, was the father of chemical thermodynamics. He applied the second law of thermodynamics and showed that law of mass action is applicable to water-gas equilibrium, determined molar volumes of liquids and measured dissociation pressures.

Cato Maximilian Guldberg (1836-1902), a mathematician published memoirs on thermodynamics and chemical equilibrium, and was instrumental in the quantitative formulation of the law of mass action in collaboration with chemist Peter Waage (1833-1900).

Henri Etienne Sainte-Claire Deville (1818-1881), a French chemist, discovered nitrogen pentoxide (N_2O_5), invented industrial process for the production of aluminum, magnesium, and sodium, worked on crystalline silicon and boron, dissociation, artificial minerals, and high pressure techniques. He also branched out in inorganic chemistry and metallurgy in collaboration with Henri Jules Debray (1827-1888), a French chemist.

Josiah Willard Gibbs (1839-1903), an American mathematical physicist, applied thermodynamics to chemistry including surface tension, thermal dissociation, and electrochemistry. He proposed the Phase Rule, developed mathematical equation that relates adsorption and interfacial tension.

Johannes Diderik van der Waals (1837-1923), a Dutch physicist, put forward the equation of state for real gases, from which the critical constants could be determined.

Thomas Andrews (1813-1885), an Irish chemist, discovered critical phenomenon and worked on thermochemistry and ozone.

Heike Kamerlingh Onnes (1853-1927), A Dutch chemist, founded Cryogenic Laboratory in Leyden, worked on critical phenomenon, low temperatures, and liquefaction of helium.

Hendrik Willem Bakhuis Roozeboom (1854-1907) was the first one to make the practical applications of Gibb's phase rule, studied triple points, chemical equilibriums, solid solutions, and used graphical methods.

Francois-Maris Raoult (1830-1901), a French chemist, laid the foundation for the modern theory of solutions started by van't Hoff by showing that depression of freezing point or of vapor pressure of the solvent by the dissolved substance (solute) could be used to calculate the molecular weight of the dissolved substance. He produced the famous law, which is known as Raoult's Law that states as "the vapor pressure of a solvent in an ideal solution decreases as its mole fraction decreases." He also studied the relation between electromotive force (EMF) of galvanic cells and heat of reaction.

Jacobus Henricus van't Hoff (1852-1911), a Dutch chemist, developed the modern theory of dilute solutions including the theory of osmatic pressure, freezing and boiling points, vapor pressure, proposed methods to determine transition points, applied the phase rule to understand the crystallization of salts from solutions. In addition, he also developed the idea of stereochemistry of carbon independently of Le Bel.

Wilhelm Ostwald (1853-1932), A German chemist, was able to spread the concepts of van't Hoff and Arrhenius on solutions through his textbooks and teachings. He improved physico-chemical methods as well as apparatus, studied the distribution of a base between two acids by refractive power and volume methods, concentrated on the rates of hydrolysis of salts and esters, the conductivities of acids, the affinity constants of acids and bases, the viscosities of solutions, and the ionization of pure water.

Johann Wilhelm Hittorf (1824-1914), a German chemist, discovered metallic phosphorus, worked on the transport of salts in electrolytic solutions, and on cathode rays.

Svante Arrhenius (1859-1927), a Swedish chemist, was the founder of theory of electrolytic dissociation, investigated the viscosity of solutions, the effect of temperature on reaction rates, was later the Director of Nobel Institute at Stockholm.

Walther Nernst (1864-1941), a German chemist, put forward the theory of galvanic cells, electrolytic solution pressure, Heat Theorem-Third Law of Thermodynamics, and atom chain-reaction theory in photochemistry. He also did research on solubility products, diffusion in solutions, liquid contact potentials, and application of quantum theory to specific heats of solids at low temperatures.

Fritz Habor (1868-1934), a German chemist, proposed the synthesis of ammonia from nitrogen and hydrogen with right combination of catalyst, temperature, and pressure – Habor process. He also studied chemical equilibrium in flames, autooxidation, electrolytic reduction of nitrobenzene, and many more electrochemical problems.

Georg Bredig (b. 1868), a German chemist, studied the dissociations of weak bases, catalytic action of colloidal platinum, the "poisoning" of catalysts, and amphoteric electrolytes.

Richard Abegg (1869-1910), a British chemist, put forward the theory of valence in association with Bodlander, distinguished between homo- and hetero-polar compounds, worked on freezing points of solutions, the dielectric constant of ice, the polyiodides of alkali metals, and potential in non-aqueous solutions.

Henry Le Chatelier (1850-1936), a French chemist, proposed the law of reaction governing the effect of pressure and temperature on equilibrium – Le Chatelier's Principle. He investigated also specific heats of gases at high temperatures, dissociation of calcium carbonate ($CaCO_3$), mass action in explosion reactions, freezing point curves, chemistry of silicates, and electrical conductivity of alloys (mixture of metals).

Sir James Walker (1863-1935), a British chemist, studied hydrolysis, acid strengths, amphoteric electrolytes, and electro synthesis of dibasic organic acids.

Sir James Dewar (1842-1923), a British chemist, invented the vacuum-jacketed vessel, liquefied hydrogen, and conducted extensive investigations on the properties of matter at low temperatures, gave the structural formula for pyridine, produced high vacuum by adsorption on charcoal cooled in liquid air, and also worked on soap films.

IIIc.c   The Rise of Inorganic Chemistry

The 19th was also an era for the development of Inorganic chemistry as well. Many inorganic chemists were also organic and physical chemists too.

Sir Henry Enfield Roscoe (1833-1915), A British chemist, investigated the effect of light on the reaction between hydrogen and chlorine that led to the foundation of quantitative photochemistry. He showed that vanadium (V) is an element related to nitrogen (N), and not to sulfur (S), its highest oxide being V_2O_5 not VO_3 as Berzelius predicted. This was an important factor later in the development of the periodic law by Mendeleev. He also made contributions to spectral analysis.

Harold Baily Dixon (1852-1930), a British chemist, worked on flame, combustion and explosions in gases. He discovered that dry mixture of carbon monoxide (CO) and oxygen (O_2) do not explode by the electric spark.

Herbert Brereton Baker (1862-1935), a Belgian chemist, studied the influence of water on chemical changes. In addition, he also invented nitrogen trioxide (N_2O_3) and the atomic weight of tellurium (Te).

Jean Servais Stas (1813-1891), a British chemist, at first worked with Dumas on organic chemistry and on the atomic weights of carbon, hydrogen, and oxygen. Later he developed very accurate methods for determining the atomic weights, which were considered to be the best for many years.

Jean Charles Galissard de Marignac (1817-1894), a Swiss chemist, discovered Ytterbium (Yb), and gadolinium (Gd), and silicotungastic acid, worked on isomorphism of salts of niobium (Nb), tin (Sn), and tungsten (W), fluorides of tin, silicon, and zirconium, and thermo-chemistry of solutions. He suggested the isotopic concepts of elements.

Sir William Crookes (1832-1919), a British chemist, discovered thallium (Tl), and studied cathode rays, rare earths, and the source of nitrogen for fertilizers.

Sir Thomas Edward Thorpe (1845-1925), a British chemist (Glasgow), discovered phosphorus pentafluoride, phosphoryl chloride, and triphosphoryl fluoride, determined the vapor density of hydrofluoric acid, obtained P_2O_4 and P_4O_6, investigated critical temperatures and viscosities of liquids. He edited *Dictionary of Applied Chemistry* and wrote the *History of Chemistry*.

Edward Divers (1837-1912), a British chemist, discovered hyponitrites, investigated ammonium carbonate, carbmate, and sulfonic acids of hydroxylamine.

Dmitri Ivanovich Mendeleev (1834-1907), a Russian chemist, put forward the periodic law. He also studied thermal expansion of liquids that led to the idea of critical temperature, worked on critical data, compressibility of gases and properties of solutions.

Bohuslar Brauner (1855-1935), a Czec chemist, carried out important investigations on chemistry of tellurium, rare earths, and periodic law.

Sir William Ramsay (1852-1916), a British chemist, began research on organic chemistry and synthesized pyridine. He then determined vapor densities of substances at various temperatures and pressures, and molecular weights of liquids, discovered inactive gases, worked on radioactivity and discovered the radium (Ra) disintegration to helium (He). Later he proposed early electronic theory of valency.

Henri Moissan (1852-1907), a French chemist, investigated fluorine compounds and metal oxides, and isolated fluorine, metal oxides, and prepared artificial diamonds. He invented electric furnace and with its help prepared metal carbides and silicon carbide.

Friedrich Raschig (1863-1928), a German chemist, was an authority on the industrial production of phenol, proposed a theory of reactions in sulfuric acid chambers, investigated sulfonic acid hydroxylamine, a technical production of hydroxylamine, nitrogen iodide, and thionic acids. He discovered chloramines and a simple method for the production of hydrazine from ammonia, and worked on industrial conversion of hydrazine into hydrazoic acid; the salt of which -lead azide - was used as a detonator in Germany during World War I. Now it is generally replaced with mercury fulminate. His contributions are distinguished by originality.

Alfred Werner (1866-1919), a Swiss chemist, was a pioneer in putting forward theories of complex compounds, which is now known as coordination chemistry, and stereochemistry of nitrogen. He introduced the important concept of coordination number; his views underlay the modern development of inorganic chemistry and paved the way to the electronic theory of valence.

Theodore William Richards (1868-1928), an American chemist, did considerable work on atomic weights, improved method for determining gravimetric atomic weights, introduced quartz apparatus, the bottling device, nephelometer, etc. He also carried out research on thermochemistry, lead isotopes, atomic volumes, and compressibilities of elements.

Alfred Stock (1876-1946), a German chemist, worked on dissociation of antimony hydride, sulfides of phosphorus, and hydrides of boron and silicon.

Otto Ruff (1871-1939), a German chemist, studied nitrogen sulfide, fluorides of metals and nonmetals, and synthesized artificial diamonds.

IIIc.d The Rise of Biochemistry

Biochemistry like other branches of chemistry also started developing in 19th century but was not considered as a separate discipline until 20th century, during which it has shaped up to be a separate branch of chemistry. Biochemistry is the interface between biology and chemistry that deals with chemical processes in the living cells.

Anselme Payen (1795-1871), a French chemist, discovered the first enzyme diastase. Later Theodor Schwann (1810-1882), a German chemist, invented another enzyme called pepsin, a digestive enzyme, thus setting the sprout of biochemistry. Luois Pasteur (1822-1895), a French chemist, demonstrated that fermentation was caused by "ferments" – later he labeled them as enzymes- in bacteria and yeast. Johann Friedrich Miescher (1844-1895), a German biochemist, discovered the nucleic acid in pus cells from patient bandages. Emil Fischer (1852-1919), a German chemist, explained the mechanism of enzyme action by lock-and-key model. Jokichi Takamine (1854-1922), a Japanese chemist, isolated the first hormone adrenaline. Eduard Buchner (1860-1927), a German biologist, discovered an enzyme zymase that causes fermentation. Arthur Harden (1865-1940), a British biologist, invented coenzymes. Phoebus Levene (1869-1940), a Russian-born American biochemist, identified the presence of ribose in RNA (Ribonucleic Acid). Frederick Banting (1891-1941), a Canadian physiologist and Charles Best (1899-1978), an American physiologist, together isolated insulin, a very important hormone that plays a crucial role in diabetes. Alexander Fleming (1881-1955), a Scottish biologist, discovered the enzyme lysozome, David Keilin (1887-1963), a Russian-born British biologist, invented cytochrome, an important enzyme in respiration of most animals and plants. James Sumner (1877-1955), an American biochemist, isolated the enzyme urease by crystallization. Hans Fischer (1881-1945), a German chemist, determined the structure of heme (the iron-porphyrin complex responsible for the red color of arterial blood) in hemoglobin. K. Lohman isolated ATP (Adenosine Triphosphate) from muscle. John Northrop (1891-1987), an American biochemist, isolated the digestive enzyme pepsin. Hugo Theorell (1903-1982), A Swedish biochemist, isolated the muscle protein myoglobin – the oxygen-transport protein of muscle.

Hans Krebs (1890-1981), a German-born British chemist, invented the mechanism for the oxidation of pyruvic acid through the tricarboxylic acid, which is known as Krebs cycle. Fritz Lipmann (1899-1986), a German-born American biochemist, identified ATP as the carrier of chemical energy in many cells. Britton Chance (1913-), an American bio-chemist, proposed enzyme-substrate complex formation to explain the action of enzymes. Alfred Hershey (1908-), an American biologist, proved that DNA (Deoxyribonucleic acid) carries genetic information. Francis Crick (1916-), a British-American biologist and James Watson (1928-), an American biologist, determined three-dimensional structure of DNA. The work of Rosalind Franklin on DNA should not be forgotten. Frederick Sanger (1918-), a British chemist and Walter Gilbert (1932-), an American molecular biologist, discovered the amino acid sequence in insulin. Arthur Kornberg (1918-), an American biochemist, invented DNA polymerase. Paul Berg (1926-), an American biologist, identified t-RNA (transfer RNA). Alick Isaacs (1921-1967), a British biochemist, dis-covered interferon – a protein capable of inhibiting the growth of infecting cells by virus. Max Perutz (1914-2001), an Austrian-born British biochemist and Sir John Kendrew (1917-1997), a British molecu-lar biologist, determined the three-dimensional structure of hemoglobin. Sydney Brenner (1927-), a South African-born British biologist and François Jacob (1920-), a French biochemist, discovered m-RNA (mes-senger RNA). Melvin Calvin (1911-1997), an American biologist, discov-ered photosynthesis and respiration in plants. Peter Mitchell (1920-1992), a British biochemist, proposed the chemiosmatic theory. Sydney Brenner and Francis Crick discovered that genetic code consists of a series of base triplets. Robert W. Holley (1922- 1993), an American chemist, Har Gobind Khorana (1922-), an Indian-born American chemist, and Marshall W. Nirenberg (1927-) , an American chemist, solved the mystery of the genetic code. Gerald Edelman (1929-), an American biochemist, analyzed immunoglobin G and determined its amino acid sequence. Howard Temin (1934-1994), an American virologist and David Baltimore (1938 -), an American biologist, discovered the enzyme reverse transcriptase. Hamilton Smith (1931 -), an American biologist, discovered restriction enzymes. Stanley Cohen (1922-), an American biologist and Herbert Boyer (1936-), an American biologist, discovered the use of restriction enzymes to produce recombinant DNA. Frederick Sanger (1918-), a British biochemist, determined the complete base sequence of DNA in becteriophase lamda X174. Alec Jeffrey, a British biochemist, devised a method to use DNA as a fingerprinting that has become very popular in crime detection. Dorothy Crawford-Hodgkin (1910-1994), a British chemist, determined the structure of vitamin B-12, penicillin, and insulin. Albert von Szent-Gyorgi (1893-1986), a Hungarian-born American chemist, studied the chemistry of muscle contractions, without his contributions the area of physical therapy would not have been developed into what it is today.

IIIc.e The Structure of an Atom and Chemical Bonding

Even though, the structure of an atom and chemical bonding are fundamental to any branch of chemistry, they closely fall in the realm of physical chemistry. Due to their importance, they deserve a special attention and hence, a brief account of the history is provided below.

In 1897 Joseph J. Thomson (1856-1940), a British physicist, showed that cathode rays discovered by Plucker, a German physicist, in 1859 were consisting of negatively charged particles known as electrons. In 1896 Henri Becqueral (1852-1908), a French Physicist, discovered radioactivity that led Pierre (1859-1906) and Madame Curie (1867-1934) to discover radium (Ra) in 1898. In 1903 Rutherford (1871-1937) and Frederick Soddy (1877-1955), a British chemist gave the explanation for radioactivity in terms of nuclear disintegration.

In 1913 independently Soddy and T.W. Richards discovered isotopes of lead from uranium and thorium. Joseph J. Thomson using the method of positive rays showed that neon is a mixture of two isotopes.

This method was extended by Aston to other non-radioactive elements and found that many of these elements were mixtures of isotopes.

In 1911 Rutherford (1871-1937), a New Zealand physicist, suggested the first atomic structure model; an atom consists of a small positively charged nucleus in the center and electrons outside of it. In 1913 Moseley determined nuclear charge by the frequency of X-rays emitted by an element. The nuclear charge is now known as an atomic number of an element that gave an element a position in the periodic table, and retains its values for different isotopes of the same element. In 1913 Neils Bohr (1885-1962), a Danish physicist, proposed a revised model of atomic structure.

In 1916 Gilbert N. Lewis (1975-1946), an American chemist, proposed the theory of chemical bonding based on the outer shell electrons. He distinguished between ionic and non-ionic (covalent) bonds: the first type is formed by the complete transfer of electrons from one atom to another atom creating a positive and negative charge atoms as explained by Kossel in 1916, which are held by the electrostatic attraction; the second type is formed by sharing of electrons in pairs – single bond is formed by one shared pair, double bond by two, and triple by three. Lewis also contributed fundamental work in chemical thermodynamics and his brilliant book *Thermodynamics* (1923) with Merle Randall (1880 – 1950) considered as one of the masterpieces in the chemistry world. In spite this, he was never awarded the Nobel Prize. Another type of bond was invented Langmuir in 1919 where both electrons in a pair come from the same atom that was further investigated by G. A. Perkins (1921); Lowery (1923) called it a "mixed double bond" or " dative bond", and Sidgwick (1923) a "coordinate link" since such bonds are present in coordination compounds. In 1931 Linus Pauling (1901-1994), an American chemist, suggested another type of bond based on the rapid shifting of electrons between bonds and he called it "resonance".

In the early part of the 20th century, the electrons were replaced by wave functions in order to account for electron configuration and chemical bonding based on the wave mechanics developed by Ervin Schrodinger (1887-1961), a Austrian physicist, based on de Broglie's (1892-1977, a French physicist) concept of dual (particle and wave) nature of an electron. Each electron was assigned a set of four quantum numbers, which arrive in the realm of quantum mechanics and the assignment was made to obey the Pauli's (1900-1958, a Austrian physicist) exclusion principle (1925). In some cases, concept of hybridization (mixing of wave functions) was introduced to account for proper valence of an atom like tetra-valence of carbon. This concept has been extended to organic chemistry in formulations of many new compounds.

The process of transmutations of elements first introduced by Rutherford in 1919 was fully exploited to synthesize new elements heavier than uranium, which are known as trans-uranium elements.

For further important developments in the later part of 20th century, see the Appendix F on Frontiers in Chemistry.

References

A Dictionay of Science, Oxford University Press, © Market House Books Ltd, 1999.

Brock, William H, *The Norton History of Chemistry*, W.W.Norton and Company, New York 1993.

Carleton S. Coon in *World Book Encyclopedia*, Field Enterprises Educational Corporation, Chicago, 1967, Vol. 15.

Maharishi Mahesh Yogi, *Bhagavad-Gita*, Penguin Books, Maryland, USA, 1969.

Partington, J.R., *A Short History of Chemistry*, Dover Publications, Inc., Westbury, NY 1989.

Salzberg Hugh W., *From Caveman to Chemist -Circumstances and Achievements*, The American Chemical Society, Washington, DC 1991.

The Facts on File Chemistry Handbook, Checkmarks Books, NY 2000.

Appendix F
Frontiers in Chemistry

I.   Organic Chemistry – General
II.   Organic Chemistry – Synthetic
III.   Organic Chemistry – Natural Products
IV.   Inorganic and Nuclear Chemistry
V.   Physical and general Chemistry
VI.   Thermodynamics
VII.   Chemical Kinetics and Chemical Changes
VIII. Nature of Chemical Bonding and Theoretical Chemistry
IX.   Polymer and Colloidal Chemistry
X.   Structural Chemistry
XI.   Analytical and Separation Chemistry
XII.   Biochemistry
XIII. Applied Chemistry
References

I. Organic Chemistry – General

1. Conformational concept -the spatial arrangement of atoms in molecules, which differ only by the orientation of chemical groups by rotation around a single bond (Sir Derek H. R. Barton (1918-1998) of London, and Odd Hassel (1897-1991) of Oslo, the Nobel Prize for Chemistry in **1969**).

2. Organometallic molecules or sandwich compounds - a metal ion is bound not to a single carbon atom but is "sandwiched"

between two aromatic organic molecules (Ernst Otto Fischer (1918-) and Sir Geoffrey Wilkinson (1921-1996), the Nobel Prize for Chemistry in **1973**).

3. Stereochemistry- not only can a compound have more than one geometric form, but chemical reactions can also have specificity in their stereochemistry. Chiral molecules - molecules that have two forms differing from one another as the right hand does from the left (Sir John Warcup Cornforth (1917-) of the University of Sussex and Vladimir Prelog (1906-1998) of ETH in Zürich, the Nobel Prize for Chemistry in **1975**).

4. Interaction of metal ions with organic molecules (Donald J. Cram (1919-) of UCLA, Jean-Marie Lehn (1939-) of Strasbourg (and Paris) and Charles J. Pedersen (1904-1989) of the Du Pont Company, the Nobel Prize for Chemistry in **1987**).

5. Carbocation (positively charged ions of hydrocarbons formed as short-lived intermediates in organic chemical reactions) Chemistry (George A. Olah (1927-) of University of Southern California, the Nobel Prize for Chemistry in **1994**).

6. Discovery of fullerenes (allotropic form of element carbon), in which 60 or 70 carbon atoms are bound together in clusters in the form of a soccer ball (Robert F. Curl, Jr., (1933-) of Rice University, Sir Harold W. Kroto (1939-) of the University of Sussex and Richard E. Smalley (1943-) of Rice University, the Nobel Prize for Chemistry in **1996**).

7. Chirally catalyzed hydrogenation reactions (William S. Knowles (1917-) of USA, Ryoji Noyori (1938-) of Japan, and K. Barry Sharpless (1941-) of USA, the Nobel Prize for Chemistry **2001**).

II. Organic Chemistry - Synthetic

1. Sugar and pine syntheses (Hermann Emil Fischer (1852-1919) of Berlin, the Nobel Prize for Chemistry in **1902**).

2. Organic dyes and hydroaromatic compounds (Adolf von Baeyer (1835- 1917) of Germany, the Nobel Prize for Chemistry in **1905**).

3. Acyclic compounds- terpenes, camphor and other compounds of ethereal oils (Otto Wallach (1837-1941) of Gottingen, the Nobel Prize for Chemistry in **1910**).

4. Pioneering work in preparative organic Chemistry - Grignard reagents and hydrogenation of organic compounds (Victor Grignard (1871-1935) in Nancy and Paul Sabatier (1854-1941) of Toulouse, the Nobel Prize for Chemistry in **1912**).

5. Elucidation of the structure of chlorophyll, the structure of hemin (the organic pigment in hemoglobin) and its synthesis from simpler organic molecules (Hans Fischer (1881-1945) of Munich, the Nobel Prize for Chemistry in **1930**).

6. Discovery of process for mass production of cortisone (Percy Julian (1899-1975) of United States of America).

7. Discovery and development of the diene (organic compounds containing two double bonds) synthesis, also called the Diels-Alder reaction (Otto Diels (1876-1954) of Kiel and Kurt Alder (1902-1958) of Cologne, the Nobel Prize for Chemistry in **1950**).

8. Total synthesis of a large number of complicated natural products, for example, cholesterol, chlorophyll and vitamin B_{12} (Robert Burns Woodward (1917-1979) of Harvard, the Nobel Prize for Chemistry in **1965**).

9. Synthetic organic Chemistry; use of boron- and phosphorus-containing compounds into important reagents in organic synthesis (Herbert C. Brown (1912-) of Purdue University and Georg Wittig (1897-1987) of Heidelberg, the Nobel Prize for Chemistry in **1979**).

10. Development of methodology for chemical synthesis on a solid matrix (Robert Bruce Merrifield (1921-) of Rockefeller University, the Nobel Prize for Chemistry in **1984**).

11. Brilliant analysis of the theory of organic synthesis; synthesis of biologically active compounds of a complexity earlier thought to be impossible (Elias James Corey (1928-) of Harvard, the Nobel Prize for Chemistry in **1990**).

III. Organic Chemistry - Natural Products

1. Structural relation between chlorophyll and hemin, magnesium as an integral component of chlorophyll, and pioneering investigations on other plant pigments, such as the carotenoids (Richard Willstätter (1872-1942) of Munich, a student of Adolf von Baeyer, the Nobel Prize for Chemistry in **1915**).

2. Structure of steroids-bile acids (Heinrich Otto Wieland (1877-1957) of Munich, the Nobel Prize for Chemistry in **1927**).

3. Structure of steroids - work on cholesterol and demonstration of the steroid nature of vitamin D (Adolf Windaus (1876-1959) of Göttingen, the Nobel Prize for Chemistry in **1928**).

4. Vitamins - synthesis of vitamin C, structure of carotene and of vitamin A. (Sir Norman Haworth (1883-1950) of Birmingham and Paul Karrer (1889-1971) of Zürich, the Nobel Prize for Chemistry in **1937**).

5. Carotenoids, isolation of vitamin B_6 and structure of vitamin B_2 (Richard Kuhn (1900-1967) of Heidelberg., the Nobel Prize for Chemistry in **1938**).

6. Sex hormones – isolation of estrone, progesterone and androsterone, and synthesis of androsterone and testosterone (Adolf Butenandt (1903-1995) of Berlin and Leopold Ruzicka (1887-1976) of ETH, the Nobel Prize for Chemistry in **1939**).

7. Plant substances – alkaloids such as morphine, synthesis of steroid hormones, and elucidation of the structure of penicillin (Sir Robert Robinson (1886-1975) of Oxford, the Nobel Prize for Chemistry in **1947**).

8. Synthesis of two hormones, vasopressin and oxytocin (Vincent du Vigneaud (1907-1997) of Cornell University, the Nobel Prize for Chemistry in **1955**).

9. Nucleotides and nucleotide co-enzymes- synthesis of ATP (adenosine triphosphate) and ADP (adenosine diphosphate), the main energy carriers in living cells, and determination of the structure of vitamin B_{12} (Alexander R. Todd (Lord Todd since 1962) (1907-1997), the Nobel Prize for Chemistry in **1957**).

IV. Inorganic and Nuclear Chemistry

1. Legand theory (Swedish chemist Wilhelm Blomstrand (1826-1862) of Lund).

2. Discovery of radioactivity (Henri Becquerel (1852-1908), Pierre (1859-1906) and Marie Curie (1867-1934) of Paris, the Nobel Prize for Physics in **1903**).

3. Isolation of element fluorine (Henri Moissan (1852-1907) of Paris, the Nobel Prize for Chemistry in **1906**).

4. Chemistry of radioactive substances and disintegration of elements (Ernest Rutherford (1871- 1937) of Manchester, the Nobel Prize for Chemistry in **1908**).

5. Discovery of the elements radium and polonium (Marie Curie, the Nobel Prize for Chemistry in **1911**, the first investigator to be awarded two Nobel Prizes).

6. Coordination theory (Alfred Werner (1866-1919) in Zürich in 1893, the Nobel Prize for Chemistry in **1913**).

7. Synthesis of ammonia from its elements, i.e., from nitrogen and hydrogen (Fritz Haber (1868-1934) of Berlin, the Nobel Prize for Chemistry in **1918**).

8. Chemistry of radioactive substances and the origin of isotopes (Frederick Soddy (1877-1956) of Oxford, the Nobel Prize for Chemistry in **1921**).

9. Isolation of heavy hydrogen (deuterium) (Harold Urey (1893-1981) of Columbia University, the Nobel Prize for Chemistry in **1934**).

10. Synthesis of new radioactive elements (Frédéric Joliot (1900-1958) and his wife Irène Joliot-Curie (1897-1956), the daughter of the Curies, the Nobel Prize for Chemistry in **1935**).

11. Isotopes as tracers involving studies in inorganic chemistry and geochemistry as well as on the metabolism in living organisms (George de Hevesy (1885-1966) of Stockholm, the Nobel Prize for Chemistry in **1943**).

12. Discovery of the fission of heavy nuclei (Otto Hahn (1879-1968) of Berlin, the Nobel Prize for Chemistry in **1944**).

13. Uranium bombardment experiments, explanation of how uranium atom was cleaved and that barium was one of the products (Lise Meitner (1878-1968), at the time a refugee of Nazism in Sweden, who had earlier worked with Hahn and taken the initiative, but no Nobel Prize was given).

14. Discoveries in the Chemistry of transuranium elements (Edwin M. McMillan (1907-1991) and Glenn T. Seaborg (1912-1999) of Berkeley, the Nobel Prize for Chemistry in **1951**).

15. Method of determining the age of various objects (geological or archeological origin) by measuring the radioactive isotope carbon-14 (Willard F. Libby (1908-1980) of the University of California, Los Angeles (UCLA), the Nobel Prize for Chemistry in **1960**).

V. Physical and General Chemistry

1. Electrolytic theory of dissociation (Svante August Arrhenius (1859-1927) of Uppsala, the Nobel Prize for Chemistry in **1903**).

2. Discovery of inert gaseous elements and their placement in the periodic table (Sir William Ramsay (1852-1916) of United Kingdom, the Nobel Prize for Chemistry in **1904**).

3. Fundamental principles governing chemical equilibria and rates of reactions (Wilhelm Ostwald (1853-1932) of Riga, the Nobel Prize for Chemistry in **1909**).

4. Accurate determinations of the atomic weights of a large number of elements (Theodore William Richards (1868-1928) of Harvard University, the Nobel Prize for Chemistry in **1914**).

5. Discovery of isotopes (atoms of the same element with the same atomic number but with different atomic masses) (Francis William Aston (1877-1945) at Cambridge University, the Nobel Prize for Chemistry in **1922**).

6. Fundamental studies of adsorption on surfaces (Irving Langmuir (1881-1957) of the research laboratory of General Electric Company, the Nobel Prize for Chemistry in **1932**).

7. Determination of electronic structure and geometry of molecules and free radicals by using molecular spectroscopy (Gerhard Herzberg (1904-1999), a physicist at the University of Saskatchewan, the Nobel Prize for Chemistry in **1971**).

8. Development of the methodology of high resolution NMR (Nuclear Magnetic Resonance) spectroscopy (Richard R. Ernst (1933-) of Zurich, the Nobel Prize for Chemistry in **1991**).

VI. Thermodynamics

1. Second Law of Thermodynamics (J. Willard Gibbs (1839-1903) at Yale, work done in **1876**).

2. Chemical thermodynamics (Jacobus Henricus Van't Hoff, The first Nobel Prize for Chemistry in **1901**).

3. Thermochemistry (Walther Hermann Nernst (1864-1941) of Berlin, the Nobel Prizes for Chemistry in **1920**).

4. Third Law of Thermodynamics (G.N. Lewis of Berkeley in **1920s**).

5. Chemical thermodynamics- behavior of substances at extremely low temperatures (William Francis Giauque (1895-1982) of Berkeley, the Nobel Prize for Chemistry in **1949**).

6. Thermodynamics of irreversible processes (Lars Onsager (1903-1976) of Yale University, the Nobel Prize for Chemistry in **1968**).

7. Non-equilibrium thermodynamics (Ilya Prigogine (1917-) in Bruxelles, the Nobel Prize for Chemistry in **1977**).

VII. Chemical Kinetics and Chemical Changes

1. Transition-state theory (Henry Eyring (1901-1982) of University of Utah never received a Nobel Prize, but Swedish Academy of Sciences in **1977** gave him its highest honor, the Berzilius Medal in gold).

2. Mechanism of chemical reactions (Sir Cyril Norman Hinshelwood (1897-1967) of Oxford and Nikolay Nikolaevich Semenov (1896-1986) of Moscow, the Nobel Prize for Chemistry in **1956**).

3. Extremely fast chemical reactions (Manfred Eigen (1927-) of Göttingen, Ronald G.W. Norrish (1897-1978) at Cambridge and George Porter (Lord Porter since 1990) (1920-) in London, the Nobel Prize for Chemistry in **1967**).

4. Mechanism of electron transfer reactions, especially in metal complexes (Henry Taube (1915-) of Stanford University, the Nobel Prize for Chemistry in **1983**).

5. Chemical kinetics – dynamics of chemical elementary processes (Dudley R. Herschbach (1932-) at Harvard University, Yuan T. Lee (1936-) of Berkeley and John C. Polanyi (1929-) of Toronto, the Nobel Prize for Chemistry in **1986**).

6. Transition states of chemical reactions using femtosecond spectroscopy (Ahmed Zewail (1946 -) of California Institute of Technology, the Nobel Prize for Chemistry in **1999**).

VIII. Nature of Chemical Bonding and Theoretical Chemistry

1. Discovery of electrons (Sir Joseph John Thomson (1956-1940) of Cambridge, the Nobel Prize for Physics in **1906**).

2. Chemical bonds – covalent (strong) bonds between atoms involve sharing a pair of two electrons (G.N Lewis (1975-1946) of United States).

3. Formulation of atomic model – electrons around the nucleus exist only in certain circular orbits (Neils Bohr (1885-1962) of Copenhagen, the Nobel Prize for physics in **1922**).

4. Investigation of Chemical bonds in **1927** (Walter Heitler (1984-1981) and Fritz London (1900-1954)).

5. Nature of chemical bond (Linus Pauling (1901-1994) at California Institute of Technology, the Nobel Prize for Chemistry in **1954**. Pauling was also awarded the Nobel Peace Prize in **1962**, making him as the only person to date to have won two unshared Nobel Prizes).

6. Molecular Orbital (MO) method and its fusion with experimental (spectroscopic) results (Robert S. Mulliken (1896-1986) of University of Chicago, the Nobel Prize for Chemistry in **1966**).

7. Theories concerning the course of chemical reactions (Kenichi Fukui (1918-1998) in Kyoto and Roald Hoffmann (1937-) of Cornell University, the Nobel Prize for Chemistry in **1981**).

8. Comprehensive theory for the rates electron-transfer reactions (Rudolph A. Marcus (1923-), the Nobel Prize for Chemistry in **1992**).

9. Density-functional theory and computational methods in quantum Chemistry (Walter Kohn (1923-) of Santa Barbara and John A. Pople (1925-) of Northwestern University, the Nobel Prize for Chemistry in **1998**).

IX. Polymer and Colloidal Chemistry

1. Discovery of process for vulcanizing rubber using sulfur (Charles Goodyear of United Sates of America (1800-1860)).

2. Discovery of nylon (Wallace Carothers of United States of America (1896-1937)).

3. Study of heterogeneous nature of colloidal solutions, namely, gold sols (aggregates of gold atoms) with ultramicroscope (Richard Zsigmondy (1865-1929) of Göttingen, the Nobel Prize for Chemistry in **1925**).

4. Construction of the utracentrifuge- disperse systems (The Svedberg of Uppsala ,the Nobel Prize for Chemistry in **1926**).

5. Discoveries in the field of macromolecular Chemistry (Hermann Staudinger (1881-1965) of Freiburg, the Nobel Prize for Chemistry in **1953**).

6. Discoveries in polymer Chemistry and technology (Karl Ziegler (1898-1973) of the Max-Planck-Institute in Mülheim and Giulio Natta (1903-1979) of Milan, the Nobel Prize for Chemistry in **1963**).

7. Fundamental theoretical as well as experimental investigations of the physical Chemistry of macromolecules (Paul J. Flory (1910-1985) of Stanford University, the Nobel Prize for Chemistry in **1974**).

8. Discovery and development of Conductive polymer (Alan H. Heeger (1936-) of University of California, Santa Barbara, Alan G, MacDiarmid (1927 -) of University of Pennsylvania, Philadelphia, and Hideki Shirakawa (1936 -) of University of Tsukuba, Japan, the Nobel Prize for Chemistry in **2000**, the last Nobel Prize of the century).

X. Structural Chemistry

1. Investigation of diffraction of X-rays (Max von Laue (1879-1960) , the Nobel Prize for Physics in **1914**).

2. Determination of crystal structure using X-rays diffraction(Sir William Bragg (1862-1942) and his son, Sir Lawrence Bragg (1890-1971), the Nobel Prize for Physics in **1915**).

3. Molecular structure through dipole moments and X-ray diffraction methods (Petrus (Peter) Debye (1984-1966), then of Berlin, the Nobel Prize for Chemistry in **1936**).

4. Structure of biological macromolecules (proteins and nucleic acids), especially, the amino-acid sequence of a protein and

insulin (Frederick Sanger (1918-) of Cambridge, the Nobel Prize for Chemistry in **1958**).

5. Structure of deoxyribonucleic acid (DNA) which controls the genetic code (Rosalind Franklin of United Kingdom (1920-1958)).

6. The first protein crystal structures (Max Perutz (1914-2001) and Sir John Kendrew (1917-1997) , the Nobel Prize for Chemistry in **1962**).

7. Discoveries concerning the molecular structure of nucleic acids (DNA) (Francis Crick (1916-), James Watson (1928-) and Maurice Wilkins (1916-), the Nobel Prize for Physiology or Medicine in **1962**).

8. Crystal structures of penicillin and vitamin B_{12} (Dorothy Crowfoot Hodgkin (1910-1994), the Nobel Prize for Chemistry in **1964**).

9. Crystallographic studies of boranes leading towards the better understanding of chemical bonding (William N. Lipscomb (1919-) of Harvard, the Nobel Prize for Chemistry in **1976**).

10. Development of crystallographic electron microscopy (Sir Aaron Klug (1926-) in Cambridge, the Nobel Prize for Chemistry in **1982**).

11. The development of direct methods for the determination of crystal structures (Herbert A. Hauptman (1917-) of Buffalo and Jerome Karle (1918-) of Washington, DC, the Nobel Prize for Chemistry in **1985**).

12. Three-dimensional structure of the photosynthetic reaction center (protein complex) (Hartmut Michel (1948-), then at the Max-Planck-Institut in Martinsried, Johann Deisenhofer (1943-) and Robert Huber (1937-), the Nobel Prize for Chemistry in **1988**).

13. Development of methods for identification and structure analyses of biological macromolecules (John B. Fenn (1917-) of USA, Koichi Tonaka (1959-) of Japan, and Kurt Wuthrich (1938-) of Switzerland, the Nobel Prize for Chemistry in **2002**).

XI. Analytical and Separation Chemistry

1. Development of organic microanalysis (Fritz Pregl (1869-1930) of Graz, the Nobel Prize for Chemistry in **1923**).
2. Electrophoresis -migration of protein molecules in an electric field (Arne Tiselius (1902-1971), the Nobel Prize for Chemistry in **1948**).
3. Invention of partition chromatography (Archer J.P. Martin (1910-) of London and Richard L.M. Synge (1914-1994) of Bucksburn (Scotland), the Nobel Prize for Chemistry in **1952**).
4. Development of polarographic methods of analysis (Jaroslav Heyrovsky (1890-1967) of Prague, the Nobel Prize for Chemistry in **1959**).

XII. Biochemistry

1. Discovery of cell-free fermentation and biochemical researches(Eduard Buchner (1860-1917) of Germany, the Nobel Prize for Chemistry in **1907**).
2. Sugar fermentation (Sir Arthur Harden (1865-1940) of London and Hans von Euler-Chelpin (1873-1964) of Stockholm, the Nobel Prize for Chemistry in **1929**).
3. Discovery that enzymes can be crystallized, preparation of enzymes and virus proteins in a pure form (James B. Sumner (1887-1955) of Cornell University, and John H. Northrop (1891-1987) together with Wendell M. Stanley (1904-1971), both of the Rockefeller Institute, the Nobel Prize for Chemistry in **1946**).
4. Photosynthesis and respiration - carbon dioxide assimilation in plants (Melvin Calvin (1911-1997) of Berkeley, the Nobel Prize for Chemistry in **1961**).
5. Discovery of sugar nucleotides and their role in the biosynthesis of carbohydrate (Luis F. Leloir (1906-1987) of Buenos Aires, the Nobel Prize for Chemistry in **1970**).
6. Fundamental work in protein Chemistry- discovery of protein assuming a specific three-dimensional structure is inherent in its

amino-acid sequence and anomalous properties of functional groups in the enzyme's active site (Christian B. Anfinsen (1916-1995) of NIH and Stanford Moore (1913-1982) and William H. Stein (1911-1980), both of Rockefeller University, the Nobel Prize for Chemistry in **1972**).

7. Photosynthesis and respiration - chemiosmotic theory (Peter Mitchell (1920-1992) of the Glynn Research Laboratories in England, the Nobel Prize for Chemistry in **1978**).

8. Recombinant DNA, i.e., a molecule containing parts of DNA from different species -methods for the determination of the base sequences of nucleic acids (Paul Berg (1926-) of Stanford, and Walter Gilbert (1932-) of Harvard and Frederick Sanger of Cambridge, the Nobel Prize for Chemistry in **1980**).

9. Discovery of the catalytic properties of RNA (Sidney Altman (1939-) of Yale and Thomas R. Cech (1947-) of the University of Colorado, the Nobel Prize for Chemistry in **1989**).

10. DNA technology - PCR (polymerase chain reaction) technique and site-directed mutagenesis (Kary B. Mullis (1944-) of La Jolla and Michael Smith (1932-) of Vancouver, the Nobel Prize for Chemistry in **1993**).

11. Elucidation of the mechanism of ATP synthesis, and discovery of an ion-transporting enzyme (Paul D. Boyer (1918-) of UCLA and John Walker (1941-) of the MRC Laboratory in Cambridge and Jens Skou (1918-) of Aarhus, the Nobel Prize for Chemistry in **1997**).

XIII. Applied Chemistry

1. Invention and development of chemical high pressure methods(Carl Bosch (1874-1940) and Friedrich Bergius (1884-1949), both of Heidelberg, the Nobel Prize for Chemistry in **1931**).

2. Agricultural and nutritional Chemistry - development of the AIV method ; fodder could be preserved with the aid of a mixture of sulfuric and nitric acid (AIV acid) (Artturi Ilmari

Virtanen (1895-1973) of Helsinki, the Nobel Prize for Chemistry in **1945**).

3. Chemical processes leading to the formation and decomposition of ozone in the atmosphere (Paul Crutzen (1933-), of the Netherlands, but working at the Max-Planck-Institute in Mainz, Mario Molina (1943-) of MIT and F. Sherwood Rowland (1927-) of Irvine, the Nobel Prize for Chemistry in **1995**).

References

Bo. G. Malmstrom, *The Nobel Prize in Chemistry: The development of Modern Chemistry*, www.nobel.se/Chemistry/articles/malmstrom/index.html, and references cited therein.

Winners of the Nobel Prize in Chemistry, *www.almaz.com/nobel/Chemistry/Chemistry.html*.

G. William Daub and William S. Seese, *Basic Chemistry*, Prentice Hall, New Jersey, 1996.

Index

Abrasive action, 3

Acetaminophen, 75, 79-82

Acetate rayon, 8

Acetic acid, 40, 80, 158, 160, 193, 197

Acetylsalicylic acid, 77

Acid rain, 20, 70

Acids, 2-3, 5, 30, 38, 40, 43, 62, 76, 78, 86, 168, 175, 178-180, 184-187, 189, 191-200, 202, 204-206, 216, 222-223, 225

  Acetylsalicylic acid, 77

  Acetic acid, 40, 80, 158, 160, 193, 197

  Ascorbic acid, 11

  Hydrochloric acid, 44, 78, 147, 158, 160, 174, 177

  Nitric acid, 159-160, 174, 177, 225

  Sulfuric acid, 2, 17-18, 20, 86, 159-160, 177-178, 198, 206

Acrylic fibers, 9

Activity, 48-49, 85, 167, 195-196

  Calisthenics, 49

  Cycling, 49

  Jogging, 49

  Running, 9, 49

  Sitting, 23, 27, 49

  Skiing, 49

  Walking, 13, 30, 49

ADP(Adenosine diphosphate), 37, 45-47, 217

Alchemy, 124, 128, 161, 164, 166, 168, 172-177, 179, 181, 195

  Chinese, 167, 169, 172, 175

  Indian, 171-172, 174-175

  Material, 8, 10, 13, 15, 29-30, 37, 39, 41, 43, 141, 145, 161, 168, 171, 182

  Medicinal, 161, 168, 176, 200

  Spiritual, 161, 164, 168, 174, 176

Algae, 50, 65

Alizarin, 39-40

Alkaline battery, 60

Akaline dry cell, 60

Alkaloid, 79

Alka-Seltzer, 78

Allotropic form, 29-30, 68, 136, 214

  Buckyball, 29-30, 93, 95, 107-108, 114, 137

  Diamond, 29, 140-141

  Graphite, 29, 60, 140-141, 143-144

Oxygen, 14-15, 20-21, 23, 25, 30, 46-48, 52, 65, 68-71, 181, 183-184, 186-188, 191, 204

Ozone, 8, 50, 68-72, 201, 226

Ammonium hydroxide, 159-160

AMP (Adenosine monophosphate), 32, 45

Amphoteric detergent, 63

Anabolic steroids, 48

Anacin, 79

Analgesics, 75-76, 78-79

Anisotropic melts, 74

Anode, 17, 19, 60

Antarctic ozone hole, 71

Anti-freeze, 22

Aristotle, 26, 169-170, 177, 181

Arrhenius, Svante, 72, 200, 203

Artificial life, 31

Aspirin, 75-82

Atomic number, 27, 210, 219

ATP (Adenosine triphosphate), 36-37, 45-47, 207-208, 217, 225

Bacteria, 2-4, 35, 49-50, 67, 207

Bases, 3, 5, 86-88, 184-187, 192, 199-200, 202-203

Ammonium hydroxide, 159-160

Sodium hydroxide, 160

Battery, 13, 16-18, 34, 59-60, 190

Bayer-timed release, 79

Big bang, 163

Binding agents, 77

Birth of chemistry, 161, 164, 180

Bleach, 64, 68

Blood clotting, 81-82

Boiling point elevation, 22

Boiling point, 22

Bond length, 95-96, 99-102, 107, 113-115, 140, 143

Briggs logarithm, 85

Buckminsterfullerene, 29, 93, 124

Buckyball, 29-30, 93, 95, 107-108, 114, 137

Buddhist elements, 170

Buffered aspirin, 77-78, 81

Bufferin, 78

Builders, 62-64

Bulk density, 136, 140

Butadiene, 19

C_{60} molecule, 93, 95-96, 98-105, 108, 110-117, 135-137, 140

Caffeine, 12-13, 42, 79

Calisthenics, 49

Calorie, 11-12

Candy bar, 42

Carbon fixation, 37

Cast (Plaster of Paris), 28

Catalytic converter, 13, 19-20, 23

Cathode, 17, 19, 60, 73, 203, 205, 209

Cationic detergent, 63

Central nervous system(CNS), 12

Cereal, 10, 12

Chalk, 26-28, 43

Chapman, 93

Chemical bonding, 162, 209-211, 213, 220, 223
Chinese alchemy, 175
Chlorine, 49-50, 64, 70-71, 177, 185-188, 190-191, 204
Chlorofluorocarbons(CFC), 69
Chromosomal damage, 42
Clausius-Clapeyron equation, 54
Cloud, 51, 53-54, 83, 108
Clusters, 135, 137, 214
Codeine, 75-76, 79, 81
Coefficient of friction, 56-57
Coefficient of viscosity, 104
Coffee, 10, 12-13
Cola drink, 42
Colloidal dispersion, 6
Colloidal solutions, 55, 221
  Dew, 54-55
  Fog, 55, 57-58
  Smog, 55, 68
Common headache, 75
Common stock solution, 155
Computer simulation, 93, 96, 135, 137
Copolymer, 10, 19
Corannulenes, 94
Cotton, 7-9, 43, 166
Crimplene, 9
Crothers, W.H., 8
CRT (Cathode Ray Tube), 73
Crystalline, 29, 74, 178, 190, 201
Cullet, 15

Cyclic-AMP, 13
Cycling, 49
Dacron, 9
Dalton, John, 26
Dark reactions, 36-37
Dead lake, 65
Delocalization energy, 116
Demineralization, 3-4
Demineralize, 2
Democritus, 26, 171
Density, 18, 31-34, 53, 108, 136, 140-141, 143-144, 151-155, 200, 205
Density method, 151, 154-155
Detergent action, 62
Detergents, 7, 44, 61-66
  Amphoteric, 63, 203-204
  Cationic, 63-64
  Neutral, 63, 78, 87-88, 108, 115-116, 136
Dew, 54-55
Dew -point, 54
Diamond, 29, 140-141
Diffusion, 56, 103-107, 200, 203
Dihedral angles, 98-99
Dilinoleo-linolenin, 38
Dipole Moment, 108, 110-111, 116-117
Disinfectant agent, 50
Dispersion energy, 137-138
Displacement technique, 33
Diuresis, 13

Dry cells, 59-60
  Alkaline dry cell, 60
  Laclanche cell, 60
  Mercury cell, 60
  Nickel cadmium cell, 60
  Wet cell, 60
Dry cleaning, 61-62, 65
Early knowledge, 161, 180
Ecotrin, 78
Effervescence, 2, 40
Effervescent-type aspirin, 77
Einstein, 57
Electrochemical theory, 185-186
Electrical energy, 16-18, 59
Electrode, 17, 57, 59-60
Electrolysis, 13, 18, 185, 193-194
Electrolytic cell, 18
Electroplating, 18
Elements, 26-31, 43, 58, 83, 167-
    171, 174, 177, 180-181, 187,
    199, 205-206, 209-211, 217-219
Elixir of life, 168, 172, 175-177
Emulsion, 6, 14, 35
Enamels, 14-15
Endergonic, 46-47
English system, 32
Enteric coated aspirin, 78
Enthalpy of formation, 112
Enzymes, 4, 46, 62, 65, 67, 78, 178,
    199, 208-209, 224
Esoteric alchemy, 161, 168, 174
Ethanol, 1, 25

Ethyl tert-butyl ether (ETBE), 25
Excedrin, 79
Exercise, 45, 47-49
Exergonic, 46-47
Exoteric alchemy, 161, 168, 174
Extracranial headaches, 75
Fabric brighteners, 65
Fabric softeners, 64
Fatty acid, 5
Fermentation, 4, 34-35, 47, 187,
    189, 195, 207, 224
Fog, 55, 57-58
Foundation of modern chemistry,
    161, 184
Four basic elements, 26
Freezing point, 22, 202-203
Freezing point depression, 22, 202
Freons, 70-72
Friction, 56-57, 67
Frictional coefficient, 103
Frozen foods, 66-67
Fullerenes, 93-94, 119-121, 144-
    145, 214
Fullerides, 94
Galvanic cells, 16, 202-203
Gasoline, 13, 23-26
Gingivitis, 1
Glass, 10, 13, 15-16, 18, 28, 67, 72,
    150-151, 164, 179
Global warming, 25
Glycerol, 5
Glycolysis, 46-47

Goodyear, Charles, 19, 221
Graphite, 29, 60, 140-141, 143-144
Greek elements, 26, 170
Greenhouse effect, 68, 72-73
Groups, 9, 11-12, 27, 45, 48, 179-180, 213, 225
Halitosis, 1
Halons, 70-71
Hard water, 4, 7, 62-63
Heat capacity, 53
Heavy ions, 7
Headaches, 41, 75-76, 79
  Intracranial, 75
  Extracranial, 75
  Migraines, 75
  Psychogenic headache, 76
  Sinus headaches, 76
Hindu elements, 170
History, 26, 40, 83, 161, 163-164, 166, 173, 195, 205, 209, 211-212
Homogeneous solution, 1
Hormones, 48, 216-217
Huckel molecular orbital method, 95, 113
Humidity, 54-55
Hydrocarbons, 20, 23-24, 65, 193-194, 196-197, 214
Hydrochloric acid, 44, 78, 147, 158, 160, 174, 177
hydrogen activity, 85
Hydrogen bonding, 9

Hydrogen bonds, 53
Hydrometer, 18
Hydrophilic, 5-6, 62
Hydrophobic, 5, 62
Hydroxyapatite, 3-4
Iatrochemy, 161, 168, 176
Ibuprofen, 75, 80-82
Icosahedron, 95
Indian alchemy, 175
Induced dipole moment, 108, 110-111
Interatomic attractive forces, 34
Interatomic repulsive forces, 34
Intermolecular separation, 135-142
Ionic radius method, 152, 154-155
Ionization potential, 108-110
Ink, 28-29, 44
Ions, 4, 7, 65-66, 77, 108, 136, 148, 151-158, 160, 214
Isotropic fluid, 75
Jogging, 49
Kilowatt-hour, 61
Kratschmer, 93, 119, 145
Krebs cycle, 47, 208
Kroto, 93, 95, 119, 122, 214
Laclanche` cell, 60
Lacquers, 15
L-ascorbic acid, 10
Latent heat of vaporization, 54
Latex paints, 14
Laundering process, 61, 65

Dry cleaning, 61-62, 65

Wet cleaning, 61-62

Le Chatelier's principle, 45, 203

Lead storage battery, 16, 18

Lennard-Jones 6-12 potential function, 137

Life, 2, 4, 6, 8, 10, 12, 14, 16, 18, 20, 22, 24, 26, 28, 30-32, 34, 36, 38, 40, 42, 44, 46, 48, 50, 52, 54, 56, 58, 60, 62, 64, 66-68, 70, 72, 74, 76, 78, 80, 82, 84, 86, 88, 90, 93-94, 96, 98, 100, 102, 104, 106, 108, 110, 112, 114, 116, 118, 120, 122, 124, 126, 128, 130, 132, 134, 136, 138, 140, 142, 144, 146, 148, 150, 152, 154, 156, 158, 160, 162, 164, 166, 168, 170, 172, 174-178, 180, 182, 184, 186, 188, 190, 192, 194, 196, 198, 200, 202, 204, 206, 208, 210, 212, 214, 216, 218, 220, 222, 224, 226, 228, 230, 232, 234, 236, 238, 240, 242

Artificial life, 31

Organic life, 31

Light reactions, 36

Linseed oil, 14, 38

Liquid crystal display, 73

Liquid crystals, 73-74

Liquid-crystal display (LCD), 73

Magnitude of frictional force, 56

Mass, 15, 27, 32-34, 57, 91, 111, 148-149, 152-154, 184, 201, 203, 215

Melting point, 10, 22, 167, 189

Mendeleev, D, 27

Mercury, 2, 5, 7-8, 10, 13, 21, 23, 26-29, 31, 33-34, 36, 38, 41-43, 45, 51, 53-54, 56, 58, 60, 66, 170, 172-173, 175-176, 181-183, 206

Mercury cell, 60

Mesolithic period, 166

Methyl tert-butyl ether (MTBE), 25

Metric system, 32

Micelles, 6, 62

Microwave oven, 66-67

Milk, 10-12, 35

Molecular point of view, 147, 160

Molecular weighing balance, 147, 149, 151

Mouthwash, 1-2

Mutans, 3

Narcotics, 76

Nematic, 74

Neutral detergent, 63

Nickel cadmium cell, 60

Nitric acid, 159-160, 174, 177, 225

Notebook, 73

Nylon, 7-9, 221

Octane, 23-24

Octane rating, 23-24

Oil paints, 14, 38, 40
Oil varnishes, 14-15
Old stone age, 165-166
Opiates, 76
Orange juice, 10
Organic life, 31
Oxidation, 19-21, 39, 50, 184, 200, 208
Oxidizing agents, 15, 43
Oxygen, 14-15, 20-21, 23, 25, 30, 46-48, 52, 65, 68-71, 181, 183-184, 186-188, 191, 204
Oxygenated gas, 25
Ozone, 8, 50, 68-72, 201, 226
Pain, 49, 75-76, 79, 81
    Somatic pain, 75
    Visceral pain, 75
Paint, 13-14, 38-40
Painting, 38-40, 83, 165
Paleolithic period, 165
Paper, 3, 28-29, 33, 43-44, 66-67
Pauling, Linus, 11, 210, 221
Pencil, 19, 28-29
Periodic table, 27-28, 210, 218
    Atomic number, 27, 210, 219
    Groups, 9, 11-12, 27, 45, 48, 179-180, 213, 225
    Periods, 27, 194
Petroleum, 23, 25
pH, 3, 38, 50, 67, 78, 85-90, 118-119, 229
Philosopher's stone, 172, 177

Phosphorylation, 46-47
Photophosphorylation, 37
Photosynthesis, 24, 36-37, 69, 208, 224-225
Pi (p) Charge, 115
Plaque, 1
Platelet aggregation, 76, 81-82
Plato, 26
pOH, 88, 90
Polarizability, 108-110, 138
Polyester, 7, 9-10
Polyhedron, 95, 119
Polymerization, 9, 194
Polymers, 8, 14-15
Polypropylene, 10
Poor visibility, 56-57
Post-alchemy, 161, 164, 166, 179
Powell's minimization technique, 98
Pre-alchemy, 161, 164-165, 169
Prostaglandin, 76, 81-82
Proteins, 11-12, 30, 35, 81, 123, 222, 224
Psychogenic headache, 76
Pycnometer, 33
Radius, 95, 99-107, 111-112, 114, 135-136, 140-142, 152-155
Rain, 1, 20, 54, 56-58, 70, 75, 166
Ramanathan, V, 72
Reactivity index, 116
Redox reactions, 17

Reduction, 19, 21, 65, 115, 117, 194, 203
Relative humidity, 55
Relaxation time, 104, 107, 111
Remineralization, 4
Renaissance in chemistry, 162, 195
Rotational kinetic energy, 111-112
Rubber, 13, 18-19, 57, 123, 146, 221
Running, 9, 49
Rust, 13, 21
Salad, 35
Salad dressing, 35
Saliva, 2-4, 180
Sand, 3, 15
Saponification, 5, 189
Scale, 7, 50, 87-88
Scattering effect, 56-57
Scattering phenomenon, 57
Scenario, 1
Scum, 7
Scurvy, 11
SI Units, 32
Significance of density, 34
Silica, 3-4, 15, 200
Sinus headaches, 76
Sitting, 23, 27, 49
Skiing, 49
Small molecules, 102
Smalley, 93, 102, 119-121, 214
Smectic, 74
Smog, 55, 68

Snickers bar, 42
Soaps, 5-7, 62, 66
Soccer ball, 30, 93, 118-119, 214
Sodium hydroxide, 160
Solar radiation, 68-69, 72
Solution, 1, 5, 8-9, 19, 22, 85, 87, 147-148, 151-155, 157, 160, 177, 186, 202-203
Sorensen, S.P.L., 85, 90-91
Spandex, 10
Spheroidal molecule, 95
Steroid nucleus, 48
Stimulant, 42
Stoke's law, 103
Streetlights, 57-58
Streptococcus, 3, 35
Stress, 45, 74-76
Styrene, 19
Styrene-butadiene rubber (SBR), 19
Sulfuric acid, 2, 17-18, 20, 86, 159-160, 177-178, 198, 206
Superacid, 88
Superbase, 88
Superdelocalizability, 116
Supersonic aircraft, 71
Surface tension, 53, 56, 62, 201
Surfactant, 62-63
Swimming pool, 49
Synthetic fiber, 8-10
Synthetic fibers, 7-8
    Crimplene, 9

Dacron, 9

Nylon, 7-9, 221

Polyester, 7, 9-10

Rayon, 8

Spandex, 10

Trevira, 9

Terylene, 9

The oxford trios, 161, 181

The radical theory, 161, 188, 190

The rise of biochemistry, 162, 206

The rise of inorganic chemistry, 162, 204

The rise of organic chemistry, 162, 195

The rise of physical chemistry, 162, 200

The structure of atom, 162, 209

The theory of phlogiston, 161, 182-185

The theory of substitution, 190

The theory of tarter, 176

The theory of valence, 161, 193, 203, 206

Thermal conductor, 21

Thermochemistry, 22, 201, 205-206, 219

Three C60 molecules (trimer), 137

Three-body interaction, 140

Thunder, 51

Toothpaste, 2-4

Toxic paints, 41

TpH, 88-89

TpOH, 88

Translational and rotational diffusion, 105

Trevira, 9

Turpentine, 38-39, 41

Two C60 molecules (dimer), 137

Two-body interaction, 140

Two-component system, 147

Tylenol, 79-80

Ultraviolet, 44, 50, 65, 68

Unleaded gas, 23

Vasoconstrictor, 42

Vedas, 167, 169, 174

Vinyl, 21, 197

Viridian, 40

Visceral pain, 75

Viscosity, 15, 53, 104, 106-107, 203

Vitamin C, 10-11, 216

Volume, 2, 25, 32-34, 72, 99-103, 114, 140, 148, 152-154, 181, 184-185, 191-192, 202

Vulcanization process, 19

Walking, 13, 30, 49

Washer protection agents, 62, 64

Water, 7, 11, 14, 20, 22-24, 26, 28, 32-33, 35-36, 43, 46, 49-57, 61-63, 65, 67-69, 72, 75, 77, 79, 87, 148, 150-156, 160, 167, 169-171, 177, 181-182, 184, 186-187, 191-192, 200, 202, 204

Watt, 61

Weight, 25, 32-33, 48-49, 51-52, 77, 82, 177, 180, 182-184, 188, 191, 202, 204

Wet cells, 60
Wet cleaning, 61-62
Yogurt, 35

About the Author

Professor Mahadev Kumbar was born and grew up in northern part of Karnatak State, India. He received his B.Sc. and M.Sc. degrees from Karnatak University, India and Ph.D. from Adelphi University, New York. He has numerous publications in computer simulations of macromolecules as well as biologically active molecules, and has written a book " Mathematical Methods for Chemistry Beginners." He has been teaching chemistry for nearly forty years and currently is an adjunct professor of chemistry at Nassau Community College, New York.

0-595-26512-X

www.ingramcontent.com/pod-product-compliance
Lightning Source LLC
Chambersburg PA
CBHW020738180526
45163CB00001B/273